MOLECULAR BIOLOGY
INTELLIGENCE
UNIT

Actin-Monomer-Binding Proteins

Pekka Lappalainen, Ph.D.
Institute of Biotechnology
University of Helsinki
Helsinki, Finland

LANDES BIOSCIENCE / EUREKAH.COM
AUSTIN, TEXAS
U.S.A.

SPRINGER SCIENCE+BUSINESS MEDIA
NEW YORK, NEW YORK
U.S.A.

Actin-Monomer-Binding Proteins

Molecular Biology Intelligence Unit

Landes Bioscience / Eurekah.com
Springer Science+Business Media, LLC

ISBN: 978-0-387-46405-3 Printed on acid-free paper.

Springer Science+Business Media, LLC, 233 Spring Street, New York, New York 10013, U.S.A.
http://www.springer.com

Please address all inquiries to the Publishers:
Landes Bioscience / Eurekah.com, 1002 West Avenue, 2nd Floor, Austin, Texas 78701, U.S.A.
Phone: 512/ 637 6050; FAX: 512/ 637 6079
http://www.eurekah.com
http://www.landesbioscience.com

Printed in the United States of America.

9 8 7 6 5 4 3 2 1

Library of Congress Cataloging-in-Publication Data

A C.I.P. Catalogue record for this book is available from the Library of Congress.

About the Editor...

PROFESSOR PEKKA LAPPALAINEN is a research director at Institute of Biotechnology, University of Helsinki, Finland. His main research interests include the small actin binding proteins twinfilin, Srv2/CAP and ADF/cofilin as well as the analysis of the roles of the actin cytoskeleton during various cellular and developmental processes. He serves on the editorial board of Cell Motility and the Cytoskeleton and is a member of Faculty of 1000 cytoskeleton section. Pekka Lappalainen carried out his PhD-thesis work at EMBL-Heidelberg, Germany and did his post-doc at University of California, Berkeley.

CONTENTS

EDITOR

Pekka Lappalainen
Institute of Biotechnology
University of Helsinki
Helsinki, Finland
Email: pekka.lappalainen@helsinki.fi
Chapter 5

CONTRIBUTORS

Christophe Ampe
Departments of Biochemistry
and Medical Protein Research
Ghent University
Gent, Belgium
Email: christophe.ampe@ugent.be
Chapter 7

Pontus Aspenström
Ludwig Institute for Cancer Research
Biomedical Center
Uppsala University
Uppsala, Sweden
Email: pontus.aspenstrom@LICR.uu.se
Chapter 9

James R. Bamburg
Department of Biochemistry
and Molecular Biology
Molecular, Cellular and Integrative
Neuroscience Program
Colorado State University
Fort Collins, Colorado, U.S.A.
Email: jbamburg@lamar.colostate.edu
Chapter 2

Marie-France Carlier
Dynamics of Cytoskeleton
and Motility Group
Laboratoire d'Enzymologie
et Biochimie Structurale
CNRS
Gif-sur-Yvette, France
Email: carlier@lebs.cnrs-gif.fr
Chapter 1

Stien Dhaese
Departments of Biochemistry
and Medical Protein Research
Ghent University
Gent, Belgium
Chapter 7

Andrea Disanza
Istituto FIRC di Oncologia Molecolare
and
Department of Experimental Oncology
Istituto Europeo di Oncologia
Milan, Italy
Chapter 8

Roberto Dominguez
Boston Biomedical Research Institute
Watertown, Massachusetts, U.S.A.
Email: rdominguez@bbri.org
Chapter 10

Emanuela Frittoli
Istituto FIRC di Oncologia Molecolare
and
Department of Experimental Oncology
Istituto Europeo di Oncologia
Milan, Italy
Chapter 8

Janel D. Funk
Department of Biochemistry
and Molecular Biology
Molecular, Cellular and Integrative
Neuroscience Program
Colorado State University
Fort Collins, Colorado, U.S.A.
Chapter 2

Bruce L. Goode
Rosenstiel Basic Medical Sciences
Research Center
Brandeis University
Waltham, Massachusetts, U.S.A.
Email: goode@brandeis.edu
Chapter 4

Ewald Hannappel
Institute of Biochemistry
University of Erlangen-Nuremberg
Erlangen, Germany
Email: ewald.hannappel@
rzmail.uni-erlangen.de
Chapter 6

Thomas Huff
Institute of Biochemistry
University of Erlangen-Nuremberg
Erlangen, Germany
Email: thymosin@biochem.uni-
erlangen.de
Chapter 6

Roger Karlsson
Department of Cell Biology
Wenner-Gren Institute
Stockholm University
Stockholm, Sweden
Email: roger.karlsson@cellbio.su.se
Chapter 3

Christophe Le Clainche
Dynamics of Cytoskeleton
and Motility Group
Laboratoire d'Enzymologie
et Biochimie Structurale
CNRS
Gif-sur-Yvette, France
Chapter 1

Uno Lindberg
Department of Cell Biology
Wenner-Gren Institute
Stockholm University
Stockholm, Sweden
Chapter 3

Elisa M. Nevalainen
Institute of Biotechnology
University of Helsinki
Helsinki, Finland
Chapter 5

Ville O. Paavilainen
Institute of Biotechnology
University of Helsinki
Helsinki, Finland
Chapter 5

Dominique Pantaloni
Dynamics of Cytoskeleton
and Motility Group
Laboratoire d'Enzymologie
et Biochimie Structurale
CNRS
Gif-sur-Yvette, France
Chapter 1

Stéphane Romero
Dynamics of Cytoskeleton
and Motility Group
Laboratoire d'Enzymologie
et Biochimie Structurale
CNRS
Gif-sur-Yvette, France
Chapter 1

Daniel Safer
Pennsylvania Muscle Institute
Department of Physiology
University of Pennsylvania
School of Medicine
Philadelphia, Pennsylvania, U.S.A.
Email: saferd@mail.med.upenn.edu
Chapter 6

Giorgio Scita
Istituto FIRC di Oncologia Molecolare
and
Department of Experimental Oncology
Istituto Europeo di Oncologia
Milan, Italy
Email: giorgio.scita@ifom-ieo-campus.it
Chapter 8

Joël Vandekerckhove
Departments of Biochemistry
and Medical Protein Research
Ghent University
Gent, Belgium
Chapter 7

Marleen Van Troys
Departments of Biochemistry
and Medical Protein Research
Ghent University
Gent, Belgium
Email: leen.vantroys@ugent.be
Chapter 7

PREFACE

Since its discovery more than 60 years ago by F.B. Straub, it has become evident that actin plays a fundamental role in a vast number of cellular processes. The regulated polymerization of actin filaments provides force for cell migration, endocytosis, phagocytosis, and a number of morphogenetic processes such as formation of axons and dendrites in neurons. Consequently, the actin cytoskeleton is also intimately involved in a variety of developmental processes in multicellular organisms. Actin is conserved among all eukaryotic organisms and recent studies have also identified actin homologues from prokaryotes. This indicates the importance of actin in the evolution of life.

Pioneering studies by Straub and co-workers demonstrated that at physiological ionic conditions actin exists mainly in filamentous form, whereas removal of salts leads to depolymerization of actin filaments into monomers. Subsequent studies by F. Oosawa and co-workers revealed that actin monomers and filaments exist in a dynamic equilibrium and provided evidence that nucleotide hydrolysis plays an important role in the monomer-to-filament transformation process of actin. More recently, studies by many laboratories around the world have revealed that the polymerization and depolymerization of actin filaments in cells is controlled by a large array of proteins that interact with monomeric and/or filamentous actin. These proteins regulate a number of different aspects of actin dynamics and cross-link actin filaments to three-dimensional structures. Much attention has been focused on the Arp2/3 complex, formins, and Spir, which nucleate the formation of new actin filaments. However, it is known that actin monomer binding proteins, which control the size, localization and nucleotide status of the unpolymerized actin pool, also play very important roles in actin dynamics and various actin-dependent cellular and developmental processes.

Recent developments in molecular biology and live-cell microscopy methods as well as in biochemical actin dynamics and motility assays have generated numerous new findings in the field of actin research. Thus, there was a need for a comprehensive up-to-date book on actin monomer binding proteins. This book summarizes the current knowledge on actin dynamics and presents examples of how actin monomer binding proteins regulate actin filament assembly and contribute to various cellular processes. It is important to note that many actin monomer binding proteins also display other activities besides interaction with unpolymerized actin. Good examples of such proteins are ADF/cofilin, which binds both monomeric and filamentous actin and promotes rapid actin filament depolymerization in cells (see Chapter 2 by Funk and Bamburg) and WASP/WAVE-family proteins, which are activators of the Arp2/3 complex (see Chapter 8 by Frittoli et al).

In the first chapter of the book Marie-France Carlier and co-workers introduce the general principles of actin dynamics and describe how various proteins can regulate the assembly and disassembly of actin filaments in cells. The following chapters present examples of different types of proteins that interact with actin monomers. Some of these are conserved in evolution from yeast to mammals (e.g., profilin, Srv2/CAP, twinfilin, verprolin/WIP; see Chapter 3 by Karlsson and Lindberg, Chapter 4 by Goode, Chapter 5 Nevalainen et al, and Chapter 9 by Aspenström), while others regulate actin dynamics in specialized cell-types of multicellular organisms (e.g., β-thymosins and ciboulot/tetrathymosin; see chapters by Hannappel et al and Van Troys et al).

Interestingly, recent structural work has revealed that most actin monomer-binding proteins bind to a specific 'hot-spot' on the surface of the actin molecule. In the last chapter of the book Roberto Dominguez summarizes the current knowledge on the mechanisms by which various actin monomer-binding proteins interact with actin.

Functions of individual actin monomer binding proteins are now rapidly being uncovered. How their activities are regulated and how these proteins in combination with each other and with other actin-binding proteins contribute to various actin-dependent cellular processes are important questions for future research. Furthermore, the details of how ATP hydrolysis affects the conformation and biochemical properties of monomeric and filamentous actin remain to be discovered.

Pekka Lappalainen, Ph.D.

CHAPTER 1

How Actin Assembly Is Modulated at Filament Barbed Ends in Motile Processes

Marie-France Carlier,* Dominique Pantaloni, Stéphane Romero and Christophe Le Clainche

Abstract

This short review is a survey of the biochemical mechanisms of control of actin filament barbed end assembly in motile processes. Regulated filament treadmilling is at the origin of barbed end growth. Barbed end nucleating, signal-responsive machineries specify the sites of filament assembly at the membrane to elicit polarized migration and determine the number of force-producing filaments. The rate of barbed end growth is controlled both by barbed end-bound factors (leaky cappers, processive motors of actin assembly) and by proteins that associate with monomeric actin and modify the rate of actin association to barbed ends. The flux of assembly at barbed ends of the different complexes of monomeric actin itself is controlled by barbed end capping proteins and by proteins that affect the rate of pointed end depolymerization, which is rate-limiting in the treadmilling cycle. While many actin-binding proteins fulfill one defined regulatory function, some of them can combine two different functions, or switch from one function to the other in a regulated fashion. Understanding the full complexity of motile behavior of living cells requires the biochemical analysis of individual actin regulatory proteins and the development of biomimetic motile systems.

Introduction

Motile processes at the origin of cell migration, cell division, morphogenesis, synaptic plasticity and endocytosis are governed by spatially and temporally controlled assembly of actin filaments. In physiological conditions actin filaments are assembled under steady-state conditions.[1-3] Actin filaments (F-actin) coexist with monomeric ATP-actin (G-actin), and turn over via pointed end depolymerization balanced by barbed end polymerization, a process known as treadmilling. Evidence coming from quantitative fluorescence speckle microscopy (qFSM) indicates that this treadmilling is modulated locally so that different populations of filaments treadmill at different rates.[4-6] We now know that both the rates of barbed end assembly and pointed end disassembly are exquisitely controlled by a variety of signaling pathways.[7,8] Thus, the stationary concentration of ATP-G-actin that is maintained in the cell at a given time results from the integrated depolymerization events from all pointed ends, which may individually depolymerize at different rates in different locations, and barbed end assembly reactions that are also spatially controlled. Barbed end growth is fed by both free ATP-G-actin and the complexes of ATP-G-actin with monomer-binding proteins that are able, like profilin-actin,

*Corresponding Author: Marie-France Carlier—Dynamics of Cytoskeleton and Motility Group, Laboratoire d'Enzymologie et Biochimie Structurale, CNRS, avenue de la Terrasse, 91198 Gif-sur-Yvette, France. Email: carlier@lebs.cnrs-gif.fr

Actin-Monomer-Binding Proteins, edited by Pekka Lappalainen.

to productively associate with barbed ends. Each of these polymerizable actin species associates to barbed ends with its own rate constant.

Site-directed barbed end assembly must occur against a membrane to generate a protrusive force, responsible for lamellipodium or filopodium extension,[2,9-11] or a pushing/pulling force that reinforces adhesion at focal contacts.[12] The forces produced depend on both the number of filaments, the structure of the actin network and the individual rates of filament assembly. These parameters are determined by membrane-bound, signal-activated machineries that nucleate new growing barbed ends using different mechanisms leading to a variety of network geometries, and by soluble actin regulatory proteins that regulate the readmilling rate. Hence different types of actin structures turnover at different rates simultaneously in the cell to achieve specific cellular function.

It is still a formidable challenge to envision how the living cell manages to orchestrate the concerted dynamics of different populations of filaments undergoing barbed end assembly in such an exquisitely specified manner. Biochemical studies simply delineate how barbed end assembly is controlled by individual proteins. Biomimetic systems simply show how the properties of a minimum number of individual proteins are integrated in a more complex modular behavior.[13-15] However so far only the living cell knows how to combine the action of different modules in a coherent fashion.

This article focusses on the recent progress made in understanding the mechanism of control of barbed end assembly within the concept of regulated filament treadmilling.

Proteins That Control Barbed End Assembly by Interacting Specifically with Actin Filament Barbed Ends

Conventional Barbed End Capping Proteins

Barbed end capping proteins are defined as proteins that bind specifically to barbed ends with a very high affinity (10^{-10} to 10^{-8} M^{-1} range), and totally block all association-dissociation events at the barbed ends. They have been among the first characterized actin-binding proteins, in the 1980's. Many strong cappers are known, the most prominent ones being members of the gelsolin family (brevin, villin, advillin, supervillin, adseverin, CapG, flightless I and FliI) that are Ca^{++}- dependent,[16,22] and heterodimeric capping protein αβ (Capping Protein, CP) homolog of sarcomeric CapZ , which are not dependent on Ca^{++}[17] and tensin a major protein from focal adhesions.[18] However the list may not be exhaustive: recently Eps8, a protein involved in Rac-dependent actin reorganization, has been shown to act by capping barbed ends;[19] vertebrate twinfilin, a known G-actin binding protein (see below), also appears to act in motile processes by capping barbed ends.[20]

These strong cappers are required for efficient motility. Blockage of barbed ends by capping proteins causes a dose-dependent increase in the steady-state concentration of ATP-G-actin, C_{SS}. In the absence of capping proteins, the value of C_{SS} is slightly higher than the critical concentration of barbed ends. Above 90% saturation of barbed ends by a capping protein, the concentration of ATP-G-actin increases appreciably[21] and at saturation it reaches the critical concentration of pointed ends, C_C^P. Treadmilling is totally abolished under these conditions. However if in this context new barbed ends are created by the action of a nucleator, these filaments grow at a rate $V_G = k_+^B .(C_C^P\text{-}C_C^B)$, for a transient period of time, until they are capped. It is only because a large number of filaments are capped that the newly created barbed ends can grow fast.

The value of C_C^P can be enhanced by proteins like ADF (see below), making the difference $C_C^P\text{-}C_C^B$ even larger. Thus capping proteins and ADF act in synergy to increase the rate of barbed end growth and enhance motility (Fig. 1).

Because they generally bind barbed ends with very high affinity, capping proteins often do not dissociate rapidly from the barbed end. The slow kinetics of interaction with the barbed ends may affect filament growth and average length.[23] Additionally, whether a given

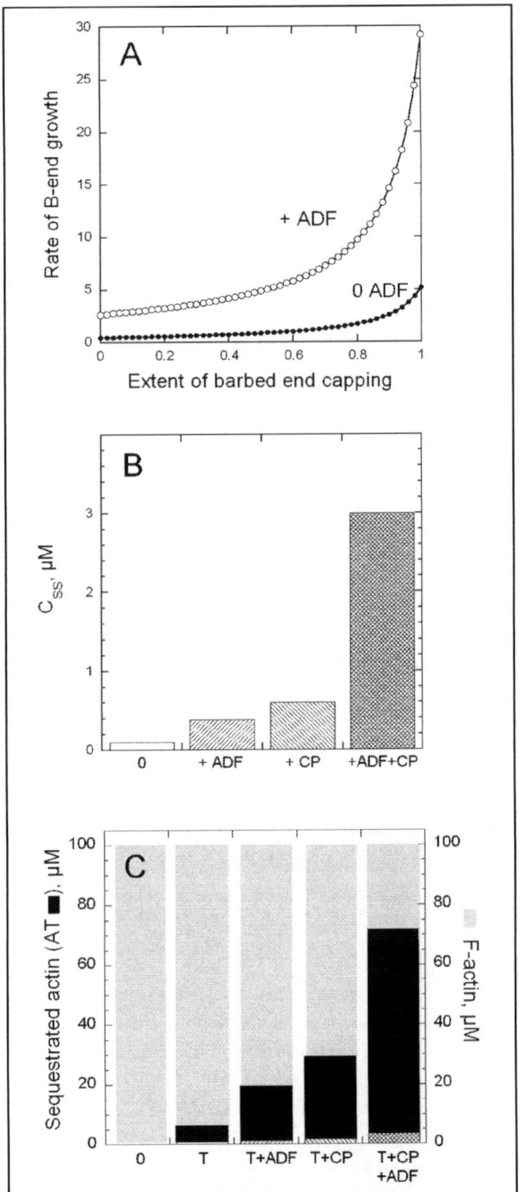

Figure 1. Combined effects of capping proteins and ADF/cofilin on barbed end growth and actin sequestration by β-thymosin. A) The rate of barbed end growth (in subunits added per second) of an uncapped filament is calculated in the absence and presence of a saturating amount of ADF, at the indicated extent of barbed end capping. B) The concentration of ATP-G-actin at steady state, Css, is calculated for pure actin in the absence of regulators (0), and in the presence of saturating amounts of ADF, Capping Protein or ADF + Capping Protein. C) Calculated steady-state amounts of free ATP-G-actin (Css, hatched boxes), Tβ4-actin (sequestered actin, black boxes) and F-actin (grey boxes), assuming total amount of actin = 100 µM, total Tβ4 = 100 µM, with K_T = 1.6 µM for the Tβ4-actin complex.

capping protein shows a preferential affinity for ADP-bound or ATP-bound barbed end terminal subunits will affect the regulation of filament growth. Under conditions of rapid growth (high C_{SS}), barbed ends have a higher probability to have ATP bound, while ADP is more frequently bound to terminal subunits at concentrations of ATP-G-actin close to C_C^B.[24] As a result, once activated, capping proteins that show higher affinity for ATP-actin at barbed ends will arrest fast barbed end growth more efficiently than slow growth. The opposite prevails for capping proteins that prefentially bind ADP-actin at barbed ends, like twinfilin.[20]

How the activity of capping proteins is regulated is partially understood. The physiological amount of these proteins is in the micromolar range, while a much lower amount is expected to be bound to barbed ends, suggesting some down regulation. PIP2 has been shown to inhibit both gelsolin and the Capping Protein.[23,25] Capping Protein is inhibited by CARMIL, a protein that not only sequesters CP but also promotes its rapid dissociation (uncapping) from barbed ends, thus creates new barbed ends rapidly, possibly in a signal-controlled fashion.[26,27] Interestingly, CARMIL may, by sequestering CP, buffer free CP at a low concentration, thus regulating the kinetics of CP association/dissociation with barbed ends and resulting barbed end growth.

Leaky Cappers

Leaky cappers are defined as capping proteins that are in rapid equilibrium with barbed ends and do not block but only modify the kinetic parameters for actin assembly-disassembly at barbed ends (k_+^B, k_-^B). Formins may act as leaky cappers on some conditions, however in living cells they more likely operate as processive motors of actin assembly (see below). So far leaky cappers are interesting theoretically, for their potential effect on actin dynamics and treadmilling. A leaky cap for instance may change the critical concentration for barbed end assembly. Typically an increase in C_C^B will lower the energetic difference between the two ends (C_C^P-C_C^B). As a result, barbed end assembly and motility will be slowed down.

Processive Motors of Actin Assembly at Barbed Ends: Formins

The dynamics of biopolymers including actin filaments, microtubules and nucleic acids are regulated by processive enzymes that catalyze several consecutive reactions of assembly or disassembly without dissociating from their substrate. The existence of a processive motor that anchors the growing barbed end of actin filaments to a surface has been hypothetized. In this model a motor is attached to a surface and binds exclusively to ATP-bound at actin barbed ends. Hence the motor uses ATP hydrolysis to slide on the growing end. This "clamped-filament elongation" would support actin-based motility.[28]

Formins are recently discovered signal-responsive proteins which nucleate actin filaments in a site-directed fashion and catalyze processive barbed end assembly in discrete motile and morphogenetic processes such as assembly of the cytokinetic ring, assembly of actin cables in yeast, filopodia extension, CR3-dependent phagocytosis, filament assembly at cadherin-dependent cell-cell contacts, and probably other still non identified actin structures involved for instance in chromatin positioning in meiosis.[29]

All formins contain Formin Homology domains FH1 and FH2 that are both required for in vivo function. The FH1-FH2 is the constitutive unit, but full length formins are in an auto-inhibited conformation, relieved by Rho GTPases.[30,31] The FH1 domain binds profilin and the FH2 domain nucleates actin filaments by stabilizing an actin dimer. Crystal structures of FH2 domains revealed a stable dimer. The FH2 core is monomeric and does not nucleate but caps filament barbed ends. A short flexible linker between FH1 and FH2 mediates dimerization and confers FH2 its nucleating activity.[32,33]

Rapid processive filament assembly from profilin-actin is catalyzed by the FH1-FH2 dimer. Kinetic studies carried out with the FH1-FH2 domain of mDia1 as a model have shown that the same formin molecule remains bound to the growing barbed end for up to 4000 consecutive G-actin association events.[34,35] Formin is not a motor itself. It uses the free energy of ATP hydrolysis associated with profilin-actin assembly to catalyze processive growth. The rate constant for assembly of profilin-actin at barbed ends is increased by one order of magnitude by end-bound formin, indicating that the assembly reaction is actually steered—electrostatically or hydrodynamically—by bound formin.[35] On the other hand, the critical concentration at barbed ends is not greatly affected by formin, which implies that the dissociation rate constant of profilin-actin from barbed ends is increased as well.

In summary formins regulate filament assembly at the barbed ends by acting on the rate parameters. This result implies that formin-based motile processes are much faster at a given concentration of ATP-G-actin, than processes mediated by free barbed end assembly. Observations made in vivo of puzzlingly fast elongation of actin cables driven by formin[36] are consistent with expectations from biochemical studies.

The effect of capping proteins on formin-based motile processes greatly differs from their effect on processes mediated by free barbed end assembly. As discussed in the previous section, capping proteins increase the energetic difference between the two ends and by this effect enhance free barbed end growth. In contrast, capping proteins compete with formins for barbed end binding and eventually cause arrest of formin-based processes.[35] Conversely, in complex motile processes like cell migration where both free barbed end assembly and formin-bound barbed end assembly are involved, depletion of capping proteins lowers the energetic difference between the two ends, thus is expected to depress free barbed end growth-mediated processes, but favor rapid persistent formin-based barbed end assembly (Fig. 2). In contrast, overexpression of capping protein is expected to arrest formin-based barbed end assembly and enhance free barbed end assembly. These effects are actually observed and play a selective role in the formation of lamellipodia and filopodia.[37] It is remarkable that the combination of simple biochemical properties allows the expression of such a large variety of motile phenotypes.

Nucleation of Barbed Ends by Filament Branching by WASP Family Proteins and the Arp2/3 Complex

WASP family proteins (WASP, N-WASP, Scar/WAVE) are signal-activated enzymes that nucleate new filaments by an autocatalytic branching process. These proteins act at the leading edge where they are targetted by a variety of signalling pathways, in different motile processes such as lamellipodium extension, phagocytosis and endocytosis.[38-39] Once activated they all use Arp2/3 complex, G-actin and a filament barbed end as substrates to catalyze the duplication of barbed ends generating a dendritic arborescent array. The growth of these filaments is transient in the cell context because it is eventually stopped by a capping protein. Under steady state conditions branched nucleation is balanced by capping, thus maintaining a stationary number of growing barbed ends. Hence the density of branching points in the array increases with the concentration of capping proteins.[10]

Coordination between capping proteins and the Arp2/3 complex is realized by the protein CARMIL. In addition of being an inhibitor of capping protreins, CARMIL is also a nucleating promoting factor that activates the Arp2/3 complex.[40] Hence CARMIL could be a critical regulator of the barbed end dynamics by coordinating two synergic activities like the inhibition of barbed end capping and the nucleation of barbed ends.

Importantly, because the WASP enzyme is immobilized, branching is a diffusion-reaction controlled process. While formins produce force by acting on the velocity of barbed end growth, WASP enzymes produce force by increasing the number of growing filaments against the membrane. The surface density and distribution of WASP at the membrane determine the morphology of the dendritic array. In this system force is produced by the growth of free barbed ends from ATP-G-actin or profilin-actin (and profilin-actin like) complexes.

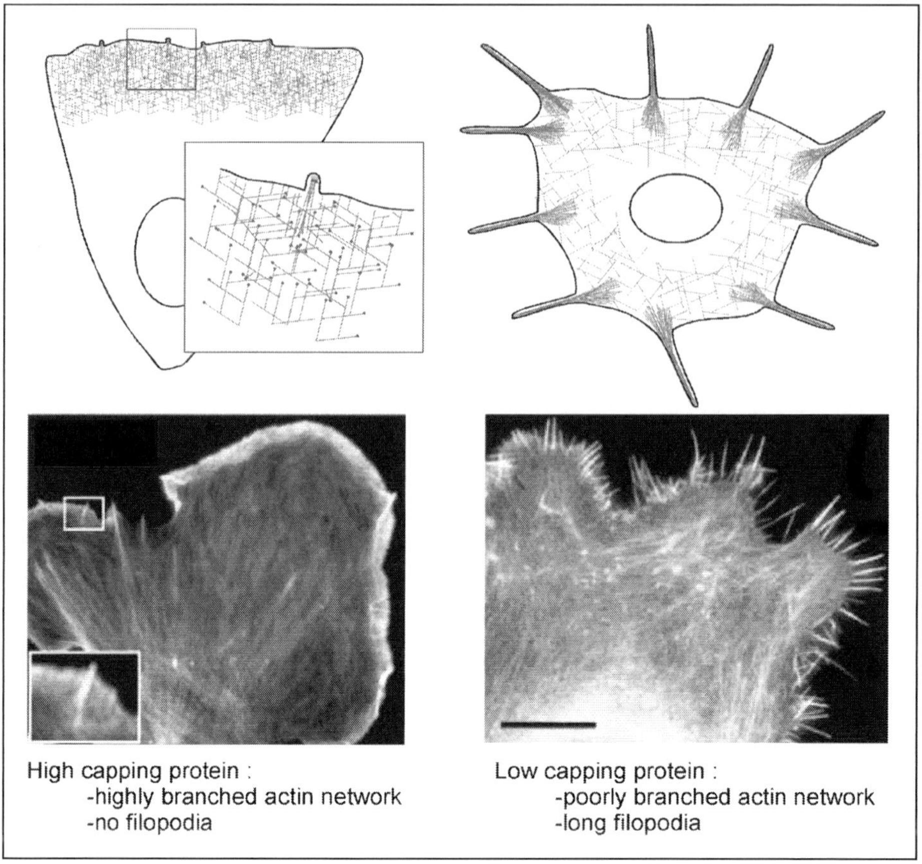

Figure 2. Capping proteins have opposite effects on N-WASP-Arp2/3 and formin-based actin assembly machineries. Actin filaments are in red, capping proteins in blue, and formin in green. Left panel: at high capping protein concentration, a densely branched filament array is formed with Arp2/3 complex: most of the filaments are rapidly capped, newly formed barbed ends grow transiently and generate efficient lamellipodia. In addition capping protein inhibits formin-catalyzed barbed end growth, thus preventing filopodia formation. Right panel: at low capping protein concentration, formin-based filament assembly is not inhibited, leading to formation of long filopodia. Incomplete capping of barbed ends leads to inefficient free filament growth in branched arrays, thus preventing lamellipodia formation. Bottom: actin staining of a B16F1 capping protein knocked down cells (right panel) and control cells (left panel). Suppression of capping protein expression leads to filopodia formation (Mejillano et al, 2005). Scale bar = 10 μm. Reproduced from: Mejillano MR et al, Cell 118(3):363-73, ©2004, with permission from Elsevier.[37] A color version of this figure is available online at http://www.Eurekah.com.

Proteins That Control Barbed End Assembly by Acting on Pointed End Disassembly: ADF/Cofilin

ADF/cofilins are known to bind G- and F-ADP-actin specifically. Activation of ADF/cofilin by dephosphorylation is induced by slingshot, which is itself activated by association with F-actin,[41] and/or by chronophin phosphatase.[42] Dephosphorylation of ADF allows its binding to ADP-F-actin, which destabilize the filaments, increasing the rate of pointed end depolymerization and the value of the pointed end critical concentration. Hence the steady-state concentration of ATP-G-actin is increased. As a consequence, the rate of barbed end growth is increased.[43] Thus it is by acting at the pointed ends that the relevant effect of ADF on lamellipodium protrusion and other motile processes is exerted. Additionally, the high value of C_{SS} favors nucleation of the few filaments that are necessary to initiate the autocatalytic barbed end branching in lamellipodia.

While these effects of ADF are well understood, how the action of ADF may be restricted to some filaments in the cell, leaving others unaffected is not understood and raises a fundamental issue. Recent studies show that in yeast ADF/cofilin participates in rapid turnover of actin filaments in actin patches but is not associated to actin cables ; similarly ADF/cofilin ensures rapid turnover of branched actin filaments in lamellipodium, but is absent from filaments in lamella which instead have tropomyosin bound.[41,44] Biochemical studies have shown that tropomyosin and ADF bind F-actin in a mutually exclusive fashion, however why some filaments bind tropomyosin and some others bind ADF/cofilin is still an open issue.[45] Structural studies show that the actin filament has a versatile structure and may exist in equilibrium with multiple conformations.[46] A given structure may easily be recognized and stabilized by appropriate ligands. Incidentally, binding of gelsolin to a filament barbed end induces a structural change that propagates over micron long distances along the polymer.[47] One may therefore speculate that the nature of a protein bound to barbed ends may induce a specific filament conformation that will favor binding of some ligands and exclude others. Such a mechanism may be at the origin of the selection of some filaments by tropomyosin and some others by ADF. In turn, the opposite effects of tropomyosin and ADF on the stability of the filament will affect the energetic difference between the two ends in opposite fashion.

Proteins That Control Filament Assembly by Interacting with G-Actin

Control of the F-Actin/G-Actin Ratio: G-Actin Sequestering Proteins

G-actin sequestering proteins (S) bind specifically G-actin, not F-actin. The major actin sequestering agents are β-thymosins which bind ATP-G-actin (A) with a 2 orders of mangnitude higher affinity than ADP-G-actin (see refs. 48,49 for a review). The SA complex is generally in rapid equilibrium with free ATP-G-actin and free sequestering protein (equilibrium dissociation constant K_S). The amount of ATP-G-actin in complex with these sequestering agents (SA) is determined by the concentration of free G-actin, which is C_{SS} when filaments coexist with G-actin, as follows.

$$[SA] = [S_0].C_{SS}/(C_{SS} + K_S)$$

As discussed above, activation of capping proteins and filament destabilizing proteins like ADF promotes the increase in C_{SS}. This results in an increase in the pool of sequestered actin, i.e., F-actin disassembly. Conversely, proteins that lower C_{SS} like tropomyosin cause F-actin assembly by decreasing the pool of sequestered actin. The rate of barbed end assembly is unaffected by sequestering proteins which by themselves do not affect motility, and simply regulate the F-actin/G-actin ratio.

At variance with β-thymosins, twinfilin is a unique example of a G-actin binding protein that sequesters ADP-G-actin specifically.[50] The ADP-actin-twinfilin complex does not polymerize and ADP does not dissociate from actin in complex with twinfilin. Hence depolymerization of pointed ends causes accumulation of ADP-actin-twinfilin in amounts that depend only on the number of filaments, not on C_{SS}. Twinfilin, as discussed above, is also a barbed end capping protein that nucleates filaments. Hence, due to its sequestering and barbed end capping activities twinfilin establishes a population of many short filaments.

Proteins That Form with G-Actin a Complex That Participates in Barbed End Growth: Profilin and the Evolution of WH2 Domains

Profilin is a clever actin-binding protein that binds to the barbed face of ATP-G-actin, thus preventing association to pointed ends, while association to a barbed end is not prevented. In addition, profilin does not appear to cap barbed ends, but in fact allows productive growth. ATP hydolysis is thought to be required to explain that the actin filament grows from profilin-actin. This view is supported by results indicating that in the presence of ADP, profilin does not support barbed end growth but only sequesters actin. The fact that profilin-actin can replace actin in maintaining the stability of filament barbed ends implies that the value of free ATP-G-actin, C_{SS}, is decreased by profilin, but the sum of C_{SS} and $[\text{profilin-actin}]_{SS}$ is conserved.[51] In conclusion, the effect of profilin is quantitatively understood in a simple system of actin + profilin + β-thymosin. Profilin acts in synergy with ADF and improves the processivity of treadmilling. In a more complex situation where both ADF and capping proteins are present together with profilin, the concentration of profilin-actin that establishes at steady state, and the resulting rate of barbed end assembly from profilin-actin in motile regions, depend on the proportion of capped filaments. To increase futher this complexity, profilin at high enough concentrations may compete with capping proteins at barbed ends.[52]

Functional homologs of profilin are represented by some members of a large family called WH2 domain proteins[53] that are characterized by the consensus actin-binding central motif LKKTET originally found in β-thymosins and variable N-terminal and C-terminal extensions. β-Thymosins are a sub-class of this family. The nature of the residues in the central and C-terminal regions determines whether the WH2 domain behaves either like a G-actin sequestering protein (like β-thymosins) or like profilin that actually promotes barbed end assembly.[54-56] In the latter category are found proteins that consist of repeated WH2 domains whose founding member is actobindin,[57] and relatives include Drosophila Ciboulot (3 repeats, see refs. 54,58) and tetrathymosin from C. elegans.[59] The crystal structure and NMR studies show that a N-terminal amphipathic helix of Ciboulot interacts with the shear zone at the barbed face of G-actin preventing association of Ciboulot-actin complex to the pointed ends. The dynamic, lose interaction of the C-terminal region of Ciboulot with the DNAse I loop of subdomain 2 of actin allows association of Ciboulot-actin to barbed ends. The recent crystal structures of WH2-actin complexes[56] fully confirm a model[54] that predicted the structure-function relationship of WH2 domains, based on the structure of Ciboulot-actin and on the sequences of various WH2 domains.

Conclusion and Perspectives

The phenomenological description of cell migration suggests that a highly complex network of regulated molecular reactions must support the coherent dynamics of actin filaments in lamellipodium, lamella, and focal adhesions during movement. On the other hand, the detailed biochemical analysis of the intrinsic properties of actin and their modulation by different regulatory proteins reveals that a wealth of diversity in the behavior of integrated systems can be generated by the combined action of only a few components. Both cell biological and biochemical studies outlined above show that in the control of barbed end growth at the origin of cell movements and changes in cell morphology, relevant parameters include: (1) the distribution of nucleating machineries in response to extracellular chemical or mechanical stimuli; (2) the number of growing filaments; (3) the extent of barbed end capping which determines the steady-state assembly flux of actin subunits; (4) the rate constants for association of barbed ends with free G-actin as well as with G-actin in complex with proteins that participate in barbed end growth. Mutual interactions between these parameters establish the functional coherence between the different actin structures involved in cell migration. In the future, further understanding of the regulation of cell motility will arise from the design of more integrated biomimetic systems challenging the principles derived from cell biological and biochemical studies.

References

1. Pantaloni D, Le Clainche C, Carlier MF. Mechanism of actin-based motility. Science 2001; 292(5521):1502-6.
2. Small JV, Stradal T, Vignal E et al. The lamellipodium: Where motility begins. Trends Cell Biol 2002; 12(3):112-20.
3. Pollard TD, Borisy GG. Cellular motility driven by assembly and disassembly of actin filaments. Cell 2003; 112(4):453-65.
4. Vallotton P, Gupton SL, Waterman-Storer CM et al. Simultaneous mapping of filamentous actin flow and turnover in migrating cells by quantitative fluorescent speckle microscopy. Proc Natl Acad Sci USA 2004; 101(26):9660-5.
5. Ponti A, Machacek M, Gupton SL et al. Two distinct actin networks drive the protrusion of migrating cells. Science 2004; 305(5691):1782-6.
6. Small JV, Resch GP.The comings and goings of actin: Coupling protrusion and retraction in cell motility. Curr Opin Cell Biol 2005; 17(5):517-23.
7. Cooper JA, Schafer DA. Control of actin assembly and disassembly at filament ends. Curr Opin Cell Biol 2000; 12(1):97-103.
8. Zigmond SH. Beginning and ending an actin filament: Control at the barbed end. Curr Top Dev Biol 2004; 63:145-88.
9. Mogilner A, Oster G. Polymer motors: Pushing out the front and pulling up the back. Curr Biol 2003; 13(18):R721-33.
10. Wiesner S, Helfer E, Didry D et al. A biomimetic motility assay provides insight into the mechanism of actin-based motility. J Cell Biol 2003; 160(3):387-98.
11. Marcy Y, Prost J, Carlier MF et al. Forces generated during actin-based propulsion: A direct measurement by micromanipulation. Proc Natl Acad Sci USA 2004; 101(16):5992-7.
12. Shemesh T, Geiger B, Bershadsky AD et al. Focal adhesions as mechanosensors: A physical mechanism. Proc Natl Acad Sci USA 2005; 102(35):12383-8.
13. Loisel TP, Boujemaa R, Pantaloni D et al. Reconstitution of actin-based motility of Listeria and Shigella using pure proteins. Nature 1999; 401(6753):613-6.
14. Upadhyaya A, van Oudenaarden A. Biomimetic systems for studying actin-based motility. Curr Biol 2003; 13(18):R734-44.
15. Carlier MF, Le Clainche C, Wiesner S et al. Actin-based motility: From molecules to movement. Bioessays 2003; 25(4):336-45.
16. Silacci P, Mazzolai L, Gauci C et al. Gelsolin superfamily proteins: Key regulators of cellular functions. Cell Mol Life Sci 2004; 61(19-20):2614-23.
17. Wear MA, Cooper JA. Capping protein: New insights into mechanism and regulation. Trends Biochem Sci 2004; 29(8):418-28.
18. Chuang JZ, Lin DC, Lin S. Molecular cloning, expression, and mapping of the high affinity actin-capping domain of chicken cardiac tensin. J Cell Biol 1995; 128(6):1095-109.
19. Disanza A, Carlier MF, Stradal TE et al. Eps8 controls actin-based motility by capping the barbed ends of actin filaments. Nat Cell Biol 2004; 6(12):1180-8.
20. Helfer E, Nevalainen E, Naumanen P et al. Mammalian twinfilin sequesters ADP-G-actin and caps filament barbed ends: Implications in motility. EMBO J 2006; 25(6):1184-95.
21. Carlier MF, Pantaloni D. Control of actin dynamics in cell motility. J Mol Biol 1997; 269(4):459-67.
22. Walsh TP, Weber A, Higgins J et al. Effect of villin on the kinetics of actin polymerization. Biochemistry 1984; 23(12):2613-21.
23. Schafer DA, Jennings PB, Cooper JA. Dynamics of capping protein and actin assembly in vitro: Uncapping barbed ends by polyphosphoinositides. J Cell Biol 1996; 135(1):169-79.
24. Carlier MF, Pantaloni D, Korn ED. Evidence for an ATP cap at the ends of actin filaments and its regulation of the F-actin steady state. J Biol Chem 1984; 259(16):9983-6.
25. Yin HL, Janmey PA. Phosphoinositide regulation of the actin cytoskeleton. Annu Rev Physiol 2003; 65:761-89, (Epub 2002 May 1. Review).
26. Yang C, Pring M, Wear MA et al. Mammalian CARMIL inhibits actin filament capping by capping protein. Dev Cell 2005; 9(2):209-21.
27. Uruno T, Remmert K, Hammer JA. CARMIL is a potent Capping Protein antagonist: Identification of a conserved CARMIL domain that inhibits the activity of the Capping Protein and uncaps capped actin filaments. J Biol Chem 2006; 281(15):10635-50.
28. Dickinson RB, Caro L, Purich DL. Force generation by cytoskeletal filament end-tracking proteins. Biophys J 2004; 87(4):2838-54.
29. Kovar DR. Molecular details of formin-mediated actin assembly. Curr Opin Cell Biol 2005.
30. Rose R, Weyand M, Lammers M et al. Structural and mechanistic insights into the interaction between Rho and mammalian Dia. Nature 2005; 435(7041):513-8.

31. Lammers M, Rose R, Scrima A et al. The regulation of mDia1 by autoinhibition and its release by Rho*GTP. EMBO J 2005; 24(23):4176-87.
32. Xu Y, Moseley JB, Sagot I et al.Crystal structures of a Formin Homology-2 domain reveal a tethered dimer architecture. Cell 2004; 116(5):711-23.
33. Otomo T, Tomchick DR, Otomo C et al. Structural basis of actin filament nucleation and processive capping by a formin homology 2 domain. Nature 2005; 433(7025):488-94.
34. Kovar DR, Pollard TD. Insertional assembly of actin filament barbed ends in association with formins produces piconewton forces. Proc Natl Acad Sci USA 2004; 101(41):14725-30.
35. Romero S, Le Clainche C, Didry D et al. Formin is a processive motor that requires profilin to accelerate actin assembly and associated ATP hydrolysis. Cell 2004; 119(3):419-29.
36. Yang HC, Pon LA. Actin cable dynamics in budding yeast. Proc Natl Acad Sci USA 2002; 99(2):751-6.
37. Mejillano MR, Kojima S, Applewhite DA et al. Lamellipodial versus filopodial mode of the actin nanomachinery: Pivotal role of the filament barbed end. Cell 2004; 118(3):363-73.
38. Miki H, Takenawa T. Regulation of actin dynamics by WASP family proteins. J Biochem (Tokyo) 2003; 134(3):309-13.
39. Innocenti M, Gerboth S, Rottner K et al. Abi1 regulates the activity of N-WASP and WAVE in distinct actin-based processes. Nat Cell Biol 2005; 7(10):969-76.
40. Jung G, Remmert K, Wu X et al. The Dictyostelium CARMIL protein links capping protein and the Arp2/3 complex to type I myosins through their SH3 domains. J Cell Biol 2001; 153(7):1479-97.
41. Sarmiere PD, Bamburg JR. Regulation of the neuronal actin cytoskeleton by ADF/cofilin. J Neurobiol 2004; 58(1):103-17.
42. Gohla A, Birkenfeld J, Bokoch GM. Chronophin, a novel HAD-type serine protein phosphatase, regulates cofilin-dependent actin dynamics. Nat Cell Biol 2005; 7(1):21-9.
43. Carlier MF, Laurent V, Santolini J et al. Actin depolymerizing factor (ADF/cofilin) enhances the rate of filament turnover: Implication in actin-based motility. J Cell Biol 1997; 136(6):1307-22.
44. Gupton SL, Anderson KL, Kole TP et al. Cell migration without a lamellipodium: Translation of actin dynamics into cell movement mediated by tropomyosin. J Cell Biol 2005; 168(4):619-31.
45. Gunning PW, Schevzov G, Kee AJ et al. Tropomyosin isoforms: Divining rods for actin cytoskeleton function. Trends Cell Biol 2005; 15(6):333-41.
46. Galkin VE, VanLoock MS, Orlova A et al. A new internal mode in F-actin helps explain the remarkable evolutionary conservation of actin's sequence and structure. Curr Biol 2002; 12(7):570-5.
47. Orlova A, Prochniewicz E, Egelman EH. Structural dynamics of F-actin: II. Cooperativity in structural transitions. J Mol Biol 1995; 245(5):598-607.
48. Carlier MF, Jean C, Rieger KJ et al. Modulation of the interaction between G-actin and thymosin beta 4 by the ATP/ADP ratio: Possible implication in the regulation of actin dynamics. Proc Natl Acad Sci USA 1993; 90(11):5034-8.
49. Paavilainen VO, Bertling E, Falck S et al. Regulation of cytoskeletal dynamics by actin-monomer-binding proteins. Trends Cell Biol 2004; 14(7):386-94.
50. Palmgren S, Ojala PJ, Wear MA et al. Interactions with PIP2, ADP-actin monomers, and capping protein regulate the activity and localization of yeast twinfilin. J Cell Biol 2001; 155(2):251-60.
51. Pantaloni D, Carlier MF. How profilin promotes actin filament assembly in the presence of thymosin beta 4. Cell 1993; 75(5):1007-14.
52. Bubb MR, Yarmola EG, Gibson BG et al. Depolymerization of actin filaments by profilin. Effects of profilin on capping protein function. J Biol Chem 2003; 278(27):24629-35.
53. Paunola E, Mattila PK, Lappalainen P. WH2 domain: A small, versatile adapter for actin monomers. FEBS Lett 2002; 513(1):92-7.
54. Hertzog M, van Heijenoort C, Didry D et al. The beta-thymosin/WH2 domain; structural basis for the switch from inhibition to promotion of actin assembly. Cell 2004; 117(5):611-23.
55. Irobi E, Aguda AH, Larsson M et al. Structural basis of actin sequestration by thymosin-beta4: Implications for WH2 proteins. EMBO J 2004; 23(18):3599-608.
56. Chereau D, Kerff F, Graceffa P et al. Actin-bound structures of Wiskott-Aldrich syndrome protein (WASP)-homology domain 2 and the implications for filament assembly. Proc Natl Acad Sci USA 2005; 102(46):16644-9.
57. Hertzog M, Yarmola EG, Didry D et al. Control of actin dynamics by proteins made of beta-thymosin repeats: The actobindin family. J Biol Chem 2002; 277(17):14786-92.
58. Boquet I, Boujemaa R, Carlier MF et al. Ciboulot regulates actin assembly during Drosophila brain metamorphosis. Cell 2000; 102(6):797-808.
59. Van Troys M, Ono K, Dewitte D et al. TetraThymosinbeta is required for actin dynamics in Caenorhabditis elegans and acts via functionally different actin-binding repeats. Mol Biol Cell 2004; 15(10):4735-48.

Chapter 2

Proteins of the Actin Depolymerizing Factor/Cofilin Family

Janel D. Funk and James R. Bamburg*

Abstract

Actin depolymerizing factor (ADF) and cofilin are the founding members of a group of structurally and functionally related actin binding proteins now collectively known as the ADF/cofilin (AC) family. AC proteins are expressed in all eukaryotic organisms, and their unique ability to bind and dynamize filamentous actin renders them essential to all cellular processes dependent upon actin dynamics. Cell division, cell motility and neuronal pathfinding, membrane dynamics, and cell polarization could not proceed without the aid of these remarkable proteins.

Overview of the ADF/Cofilin Family

ADF and cofilin proteins were initially identified and named according to their ability to either depolymerize filamentous actin (ADF) or form cofilamentous structures with actin (cofilin). Chick brain ADF and porcine brain cofilin are the founding members of the ADF/cofilin family, each discovered due to its namesake activity.[1-3] The discovery of starfish depactin,[4] bovine brain ADF,[5] porcine brain and kidney ADF (aka destrin),[6,7] and *Acanthamoeba castellanii* actophorin soon followed,[8] and with the advent of cloning and cDNA sequencing there have been numerous additions to the ADF/cofilin family based on sequence homology. A partial list of ADF and cofilin proteins found in various organisms is shown in Table 1. Because there are no clearly distinctive biochemical properties that distinguish an ADF from a cofilin, new members are usually designated as ADF or cofilin depending on the sequence to which the new gene is most closely related. Many vertebrates have genes encoding one ADF and two cofilins, and within a single organism ADF and cofilin share about 70% sequence identity (see human ADF and cofilin sequence alignments in Fig. 1A). Many lower eukaryotes such as the fruit fly (*Drosophila melanogaster*), starfish (*Asterias amurensis*) and yeast (*Saccharomyces cerevisiae* and *Schizosaccharomyces pombe*) possess only one AC family member. Nevertheless, the functional similarity of ADF to cofilin is evident in that yeast mutants lacking cofilin can be rescued with either mammalian cofilin or ADF,[9,10] and cell division and motility defects due to knockdown of cofilin-1 or ADF in mammalian nonmuscle cells can be rescued by overexpression of the remaining isoform.[11] A comparison of protein sequences and three-dimensional structures of ADF and cofilin proteins found in organisms as diverse as mammals, protozoans, yeast and plants reveals a high degree of conservation among members of this family across phylogeny (Fig. 2). Indeed, as their sequence homology suggests, ADF and cofilin have very similar activities. Both proteins cosediment with actin at pH <7.1, and both proteins depolymerize F-actin,

*Corresponding Author: James R. Bamburg—Department of Biochemistry and Molecular Biology, Colorado State University, Fort Collins, Colorado 80523, U.S.A.
Email: jbamburg@lamar.colostate.edu

Table 1. Isoforms of ADF and cofilin found in various organisms

Organism	Species	Known AC Isoforms[A,B]
Mammals	*Homo sapiens (human)*	**ADF***, Cofilin-1, Cofilin-2
	Sus scrofa (pig)	ADF, Cofilin-1
	Rattus norvegicus (rat)	Destrin/ADF,[A] Cofilin-1
	Mus musculus (mouse)	**ADF,** Cofilin-1, Cofilin-2
Chicken	*Gallus gallus*	**ADF**, Cofilin-2
Frog	*Xenopus laevis*	**XAC-1**, XAC-2
Nematode	*Caenorhabditis elegans*	Unc60A, **Unc60B**
Fruit fly	*Drosophila melanogaster*	**Twinstar**
Starfish	*Asterias amurensis*	Depactin
Sea slug	*Aplysia kurodai*	Cofilin[B]
Yeast	*Saccharomyces cerevisiae*	**Cofilin**
	Schizosaccharomyces pombe	**Adf1/Cofilin**[A,B]
Protozoa	*Dictyostelium discoideum (slime mold)*	**Cofilin-1**, Cofilin-2
	Acanthamoeba castellanii (amoeba)	**Actophorin**
	Toxoplasma gondii	**ADF**
	Plasmodium falciparum	ADF-1,[B] ADF-2
	Entamoeba histolytica	**Actophorin**[A]
Plants	*Arabidopsis thaliana*	**ADF-1**, ADF-2, ADF-3, ADF-4, ADF-5, ADF-6, ADF-7, ADF-8, ADF-9, ADF-10
	Zea mays (corn)	**ADF-1**, ADF-2, ADF-3
	Lilium longiflorum (lily)	**ADF**
	Oryza sativa (rice)	**ADF-1**, ADF-2
	Brassica napus (rapeseed)	**ADF**

This table constitutes a partial listing of AC isoforms found in various organisms, designated as ADF, Cofilin, *etc.* according to reference 141 unless otherwise noted (see footnotes A and B). * AC isoforms in bold type are also featured in Figure 2 (sequence alignments of AC proteins across phylogeny). [A] Isoforms gathered from and named according to NCBI protein database: *R. norvegicus* destrin/ADF (NP_001028838), S. *pombe* cofilin (P78929; also known as Adf1[97]) and *E. histolytica* actophorin (EAL46302). [B] Isoforms gathered from and named according to published literature: *A. kurodai* cofilin,[142] *D. discoideum* cofilin-2,[143] and *P. falciparum* ADF-1 and ADF-2.[38]

albeit to significantly different degrees, at pH >7.3. Recent literature has thus considered these two proteins to be isoforms with similar activities, and in the majority of our discussion we will describe the activities of ADF and cofilin together, referring to them as "AC proteins."

Despite the similarities between ADF and cofilin, these two proteins do exhibit some quantitative differences in their in vitro behavior and qualitative differences in their regulatory mechanisms that certainly merit discussion and careful consideration. For instance, an increase in the actin monomer pool in cells causes a robust increase in the phosphorylation of both ADF and cofilin, however ADF expression subsequently drops while cofilin expression remains unchanged.[12] An even more dramatic example of the differences between these proteins has been demonstrated in mice, as the developmental consequences of knocking out the cofilin-1 gene hardly mimic the phenotype of mice lacking the ADF gene. Whereas cofilin-1 knockout mice are embryonic lethal at day 10.5 due to severe defects in neural crest cell migration and a complete lack of neural tube closure,[13] mice lacking ADF are born healthy, but develop irregular thickening of the corneal epithelium and are blind by four weeks of age.[14] In both of these cases, the remaining AC isoform was found to be upregulated in the affected tissue, however it is evident that this compensation was not adequate for a full rescue in all tissues. Thus in

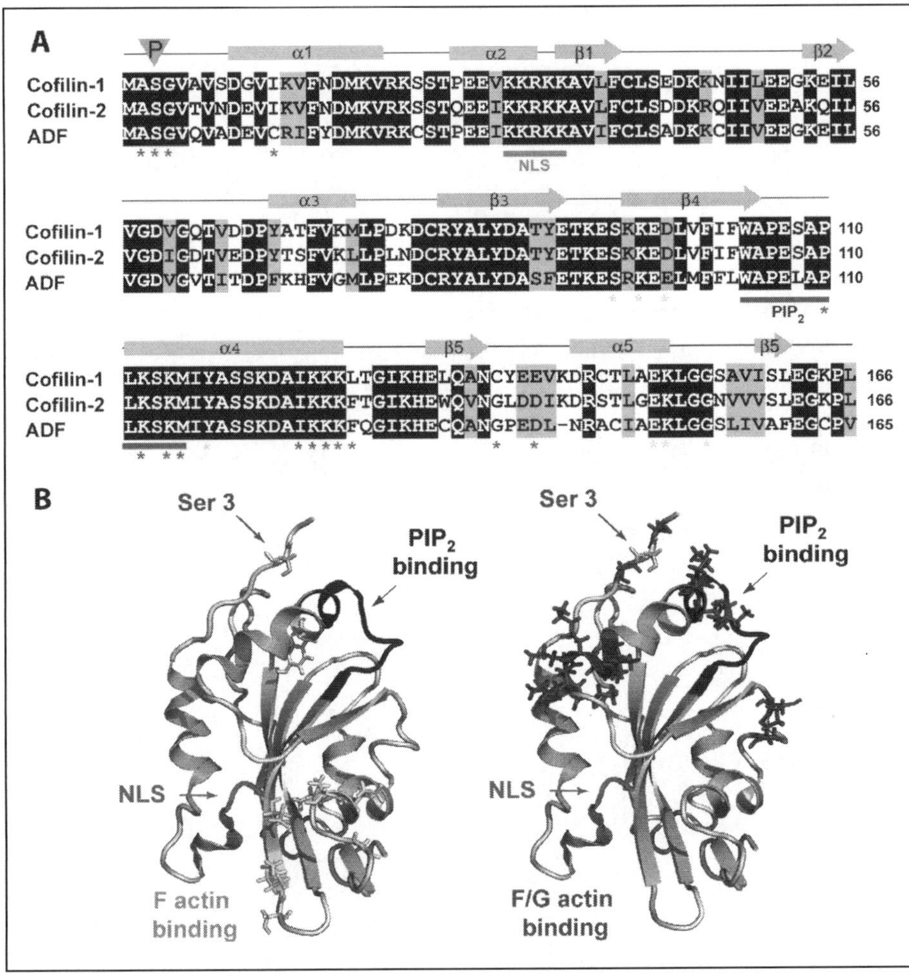

Figure 1. A) Human ADF and cofilin sequences are highly conserved. B) Structure of human cofilin-1 showing two distinct actin binding surfaces. The known human AC isoforms cofilin-1, cofilin-2 and ADF are aligned with conserved residues shaded black and similar residues shaded gray. A representation of secondary structural elements is shown above the sequences with colors matching in the three-dimensional structures shown (human cofilin-1, PDB: 1Q8X).[147] Residues essential for F- and G-actin binding are marked with a red asterisk underneath the sequence and red coloring in the right-hand structure. Residues essential for F-actin binding only are marked with a yellow asterisk underneath the sequence and yellow coloring in the left-hand structure. These designations are based on experiments performed by Lappalainen et al (ref. 16). The PIP_2 binding region is shown in blue, and the nuclear localization sequence (NLS) is shown in green. The sequences used in the alignments were aquired from the NCBI protein database and are as follows: *Homo sapiens* cofilin-1 (NP_005498), *Homo sapiens* cofilin-2 (NP_619579) and *Homo sapiens* ADF (AAB28361). A color version of this figure is available online at www.Eurekah.com.

organisms and cells where both ADF and cofilin are expressed, there are overlapping as well as distinct roles for each isoform. The intriguing interplay between ADF and cofilin isoforms has yet to be fully understood and warrants much future study.

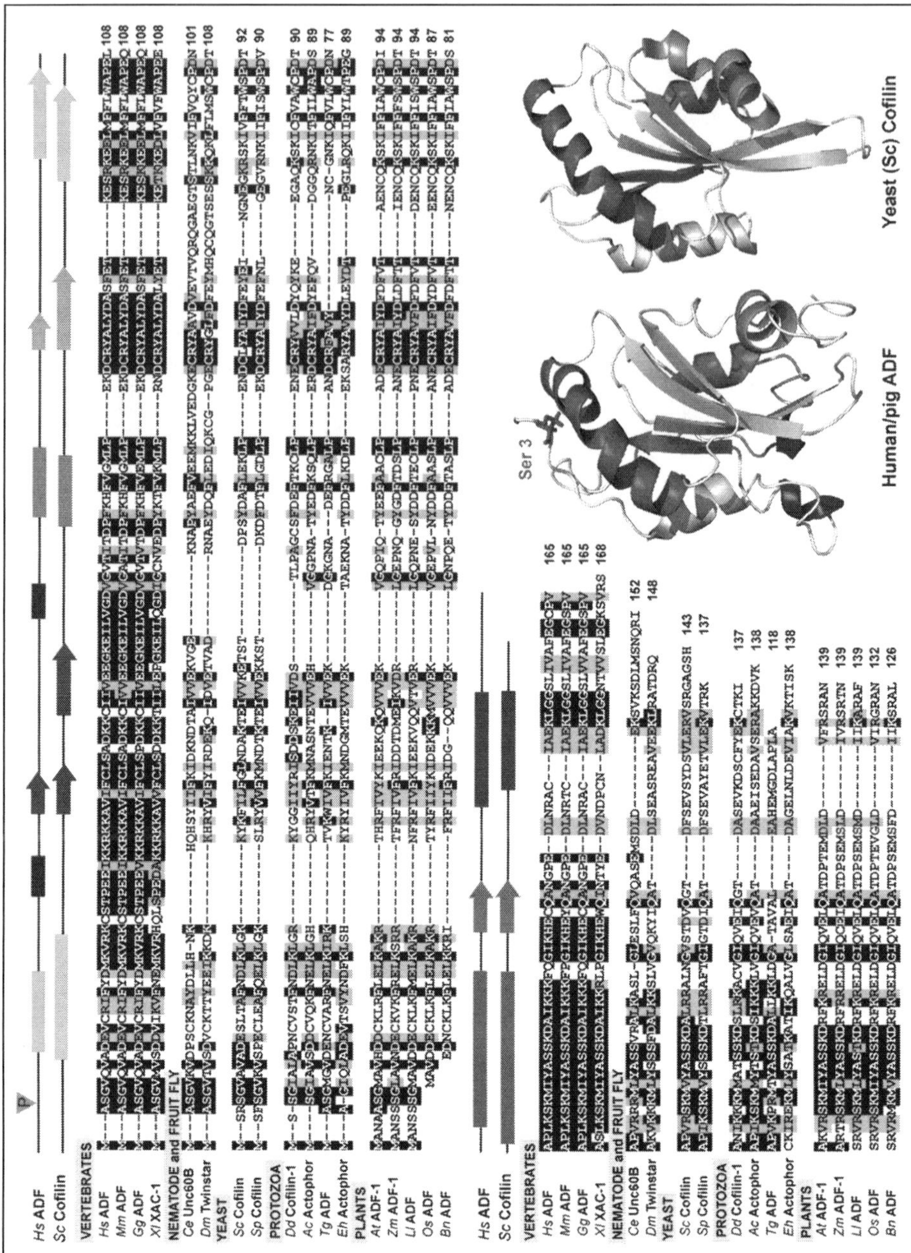

Figure 2. Sequence alignment and three-dimensional structures of AC proteins across phylogeny. AC proteins found in various organisms are aligned according to the structure-based alignment proposed by Bowman et al (ref. 144) with some additional manual adjustments. Black shading indicates residues conserved in most vertebrates; gray shading indicates residues similar in most vertebrates. Secondary structure designations are also in accordance with Bowman et al (ref. 144). Three-dimensional structures shown are porcine ADF/destrin (PDB: 1AK6)[145] and yeast cofilin (PDB: 1QPV).[146] Figure legend is continued on next page.

Figure 2, continued. The sequence of porcine ADF is identical to that of human ADF. Figures were generated using PyMOL (Copyright© 2006 DeLano Scientific LLC). Colors of alpha-helices and beta-sheets in each structure correspond to the secondary structure diagrams located above the sequence alignment. The conserved N-terminal serine is also shown in pink in the human/pig ADF structure, however N-terminal residues 1-5 and C-terminal residues 141-143 in yeast cofilin are missing due to disordering of those residues in the crystallized protein. Sequences used in the alignment are as follows: *Homo sapiens* ADF (NCBI protein database ID no. AAB28361); *Mus musculus* ADF (Q9R0P5); *Gallus gallus* ADF (AAA48575); *Xenopus laevis* XAC-1 (AAH44691); *Caenorhabditis elegans* unc60B (Q07749); *Drosophila melanogaster* twinstar (P45594); *Saccharomyces cerevisiae* cofilin (AAA13256); *Schizosaccharomyces pombe* cofilin (P78929); *Dictyostelium discoideum* cofilin-1 (XP_642259); *Acanthamoeba castellanii* actophorin (P37167); *Toxoplasma gondii* ADF (AAC47717); *Entamoeba histolytica* actophorin (XP_651689); *Arabidopsis thaliana* ADF-1 (NP_190187); *Zea mays* ADF-1 (P46251); *Lilium longiflorum* ADF (P30175); *Oryza sativa* ADF (XP_475079); *Brassica napus* ADF (CAA78482). A color version of this figure is available online at www.Eurekah.com.

Structure and Function of ADF/Cofilin

Protein Structure

All members of the AC family are 118 -168 amino acids in length (13-19 kDa) and contain a highly conserved folded domain called the ADF homology (ADF-H) domain.[15] This domain consists of a four stranded mixed β-sheet sandwiched between two pairs of α-helices, and serves as the actin-binding module (Figs. 1B and 2). Loop regions connecting secondary structure elements are less conserved. In the smallest member of the AC family, *Toxoplasma gondii* ADF (118 amino acids), the ADF-H domain is present with no extra amino acids in the loops connecting β-strands and α-helices (note the numerous gaps in the TgADF sequence shown in Fig. 2).

The actin-binding residues of AC are organized such that they form two distinct actin-binding surfaces.[16] When AC binds to filamentous actin (F-actin), it contacts at least two different actin subunits that most likely bind AC along these surfaces: one actin subunit contacting each surface. In Figure 1, residues essential for binding F-actin as well as actin monomers (G-actin) are shown in red. Residues required only for F-actin binding are shown in yellow. An inositol phosphate (PIP_2) binding site (blue) lies next to the essential F/G-actin binding surface, thus inhibiting AC binding to actin.[17] Also influencing actin-binding is a single regulatory phosphorylation site near the N-terminus (pink), conserved in most AC proteins.[18] One final important structural feature of AC proteins is a nuclear localization sequence (NLS, green), accessible when AC is bound to G-actin, but not readily accessible when AC is bound to F-actin.[19-21] The physiological function of this sequence will be discussed in more detail later.

ADF/Cofilin Function in Regulating Actin Dynamics

True to their names, ADF and cofilin proteins play a critical role in binding to and dynamizing actin filaments. Actin exists in cells as monomers or filaments and can rapidly convert between these forms with the aid of actin-binding proteins such as AC (Fig. 3). In the basic mechanism of actin dynamics, actin monomers with bound ATP (ATP-G-actin) assemble into F-actin polymers, with hydrolysis of ATP rapidly following assembly and loss of inorganic phosphate (Pi) lagging behind. The rate of subunit addition to the F-actin is largely determined by the local monomer concentration. There are unequal critical monomer concentrations for assembly at the two filament ends, therefore within a particular concentration range of actin monomer, filaments are capable of undergoing a process called "treadmilling," wherein ATP-G-actin adds to the faster growing plus end (also known as the barbed end from decoration of filaments with fragments of myosin that form an arrowhead structure) while ADP-G-actin is lost at the opposing minus end. ADP must then be exchanged for ATP on G-actin before the monomer can be reassembled into a polymer. pH, the ionic environment, and several actin-binding proteins all cooperate in regulating actin assembly and disassembly for accomplishing various biological processes.

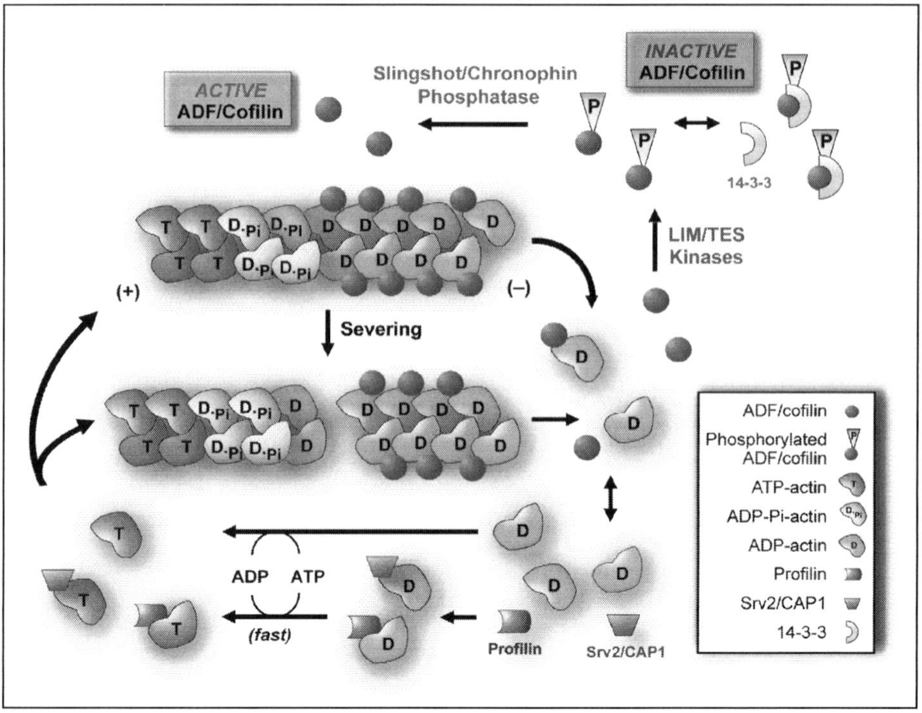

Figure 3. ADF and cofilin enhance actin dynamics. Binding of ADF/cofilin proteins to filamentous ADP-actin enhances the off rate from the minus (-) end and may lead to severing of the filament. LIM or TES kinases phosphorylate and inactivate ADF/cofilin; slingshot or chronophin phosphatases remove the phosphate and reactivates ADF/cofilin. 14-3-3 binds to phosphorylated ADF/cofilin and limits the accessibility of some phosphatases, restricting dephosphorylation. Profilin and Srv2/CAP1 bind to ADP-actin monomers and enhance the rate of ADP/ATP exchange. ATP-actin monomers are then free to be added to the plus (+) end of existing filaments. A color version of this figure is available online at www.Eurekah.com.

Proteins in the ADF/cofilin (AC) family are considered to be the primary factors in generating high turnover rates of actin filaments in vivo. This designation is based on the ability of AC to bind to ADP-F-actin and stabilize a naturally twisted minor conformer of F-actin.[22,23] AC binding weakens the longitudinal interactions between actin protomers,[24] which is the likely cause of the observed 5 to 20-fold increase in minus end depolymerizing rate. AC bound to ADP-G-actin is released from the pointed end of the filament, and nucleotide exchange is inhibited as long as the AC remains bound.[25] The affinity of some ACs for ADP-actin is up to 60 times greater than their affinity for ATP-actin,[26] thus AC slows the rate of recycling actin monomers back to the assembly competent ATP-actin pool.

Another mechanism by which AC generates high filament turnover rates is its filament severing activity. AC binds to F-actin in a highly cooperative manner[27,28] and ultimately leads to severing of the filament in one or more places. The determinants of severing sites are not completely understood: AC-binding may induce breakage at or near its own contact points, or breakage may occur within unbound sections of filaments that lie between twisted, AC-saturated regions.[29] Either way, severing contributes to increased filament turnover rates as there is an increased number of pointed ends from which subunit dissociation can occur (reviewed in refs. 30, 31) as well as new free barbed ends for actin assembly.[32-34]

ADF/Cofilins as Actin Monomer Binding Proteins

Although the dynamizing effect of AC proteins on actin filaments appears to be their major cellular function, it is also important to mention the impact ACs may have on cell biology simply as a monomer binding proteins. Developing chick brain cells possess a large G-actin pool, maintained primarily by the actin sequestering protein thymosin β4 (Tβ4). ADF and cofilin are also present in these cells, and chick ADF (but not cofilin) has a relatively strong affinity for ATP-G-actin.[35] ADF can thus work alongside Tβ4 as an important modulator of the G-actin pool.[36] ADF and cofilin, along with Tβ4 and profilin (another actin binding protein discussed later) are also known to play a critical role in maintaining the G-actin pool in developing and degenerating embryonic chick muscle.[37] This monomer sequestering ability of AC proteins is directly related to their differences in binding affinities for ATP-G-actin versus ADP-G-actin,[26] which vary from about 1:4 for chick ADF to >1:40 for chick cofilin.[35] In the parasite *Plasmodium falciparum*, PfADF-1 (one of two known AC homologues in this organism) binds exclusively to monomeric actin, preferring to interact with ADP-G-actin.[38] In fact, PfADF-1 seems to behave more like profilin or Srv2/CAP, enhancing ADP/ATP exchange on the monomer and thus increasing the pool of polymerization-ready ATP-G-actin. This is a mechanism that works well for this parasite, as *Plasmodium* F-actin is intrinsically unstable.[39] These characteristics of PfADF-1 render it rather unique among the AC protein family, however its impact on actin dynamics solely as a monomer binding protein informs us that there is indeed more to AC proteins than just their F-actin severing/depolymerizing activity.

Regulation of ADF/Cofilin Activity

Proper maintenance of the actin cytoskeleton, be it stable or dynamic, is essential to the viability of all cells. In some cell types, such as neurons and migratory cells of the immune system, the need for precise timing and directionality in actin filament rearrangements to achieve cell movement is obvious, and also impressive. It is not at all surprising that AC activity is held tightly in check with several overlapping regulatory mechanisms. A key determinant of AC activity is its state of phosphorylation, however isoform expression, compartmentalization, pH, a handful of interacting molecules and proteins, and the presence of competing actin-binding proteins in the cell all influence AC activity to some degree.

Specific Isoforms

Metazoans express multiple isoforms of ADF and cofilin in a developmental- and tissue-dependent manner. This is primarily achieved through gene duplication, however in the nematode *Caenorhabditis elegans*, at least two AC isoforms arise from a single gene through alternative splicing.[40] Although the gene products, unc60A and unc60B, display 38% sequence identity, these AC homologues differentially regulate actin dynamics, suggesting that the two proteins have separate functions in vivo. Indeed, unc60B is muscle specific whereas unc60A is ubiquitous.

Vertebrates also express a specific isoform of AC in developing muscle, called m-cofilin (cofilin-2).[41] In embryonic muscle, ADF (destrin) and nonmuscle cofilin (cofilin-1) are present in substantial amounts.[37,42-44] ADF levels decline rapidly during myogenesis in vivo. The nonmuscle isoform of cofilin is replaced by m-cofilin during myogenesis in vivo and in vitro.[41] In dystrophic and denervated mature muscle, ADF expression remains below detectable levels, but cofilin expression is upregulated probably due, at least in part, to the increase in regenerating muscle cells.[37,45,46]

ADF is expressed at low levels in mice during early embryonic development, but is postnatally upregulated in epithelial rich tissues such as the stomach, intestine and retinal pigmented epithelium. Cofilin-1 is expressed ubiquitously but during development it is highest in limb buds, somites and the neural tube.[13]

Three isoforms of plant ADF, designated ZmADF-1, 2, and 3, have been identified in *Zea mays*. The mRNAs encoding ZmADF-1 and ZmADF-2 are specific to mature pollen grains and germinated pollen, whereas ZmADF-3 is found in all other vegetative tissues examined (leaf, shoot, root, cob, and embryo).[47] ZmADF-1 is found in extending pollen tubes, and ZmADF-3 is involved in tip growth in root hair cells. These are somewhat similar processes that involve the rapid delivery of vesicles to the growing tip (reviewed in ref. 48).

Nuclear Localization and Rod Formation

Under conditions of ATP-depletion, heat shock, dimethyl sulfoxide (DMSO) treatment or high G-actin concentration in the cytosol, cofilin has been shown to translocate with actin to the nucleus,[20,21,49,50] where they form rod-like structures.[50] Similar stressors also induce AC-actin rod formation in axons and dendrites of cultured hippocampal neurons, sometimes spanning the diameter of the neurite, disrupting the microtubule cytoskeleton, and ultimately leading to degeneration of the neurite segment distal to the rod.[51] The physiological function of these rods remains unclear, however it has been demonstrated that nuclear translocation or aggregation of AC into rods is a cellular energy conservation mechanism:[51a] actin dynamics, which can utilize up to 50% of cellular ATP,[52] are slowed until the stressors are removed.[51] Resolving the mysteries of rod formation and AC nuclear localization are sure to occupy researchers for some time to come.

Phosphorylation

Actin depolymerization slows dramatically upon phosphorylation of AC.[53] An N-terminal serine phosphorylation site is conserved across phylogeny (Ser 3 in animals and insects,[18,54-57] Ser 2 in amoeba[58] and Ser 6 in plants[59]) although phosphorylation has not yet been demonstrated in some organisms. The observed slow down in actin dynamics upon phosphorylation of AC is due to a 20-30 fold decrease in its affinity for actin.[18,53,58,59] As the three-dimensional structure of the protein does not appear to change upon phosphorylation,[58] the most plausible explanation for the change in affinity is that the negative phosphate repels the positive AC-binding interface of actin.

Two related families of ubiquitous kinases, LIM and TES kinases, phosphorylate ACs. **L**in-11, **I**sl-1 and **M**ec-3 kinase (LIMK), a serine/threonine kinase containing LIM and PDZ domains, phosphorylates vertebrate ACs in vitro and in vivo.[60,61] The LIM domain targets the kinase to the Golgi, whereas the PDZ domain targets the kinase to the plasma membrane or neuronal growth cone.[62] There are two LIMK isoforms in vertebrates: LIMK1 (ubiquitous, but enriched in axonal and dendritic growth cones)[60,61,63] and LIMK2 (ubiquitous),[64-66] both regulated downstream of the Rho family of small GTPases.[67-71] LIM kinases form homodimers[72] that transphosphorylate each other upon phosphorylation of Thr 508 in LIMK1 and Thr 505 in LIMK2[70,73] by PAK or Rho-associated (ROCK) kinases, thus attaining full activation.[68,72]

Testicular protein kinase 1 (TESK1), originally discovered in testicular germ cells of rats and humans, is a serine/threonine kinase with a structure composed of a kinase domain related to those of LIM kinases and a unique C-terminal proline-rich domain.[74,75] Like LIM kinases, TESK1 phosphorylates vertebrate ACs specifically at Ser 3, both in vitro and in vivo.[76] TESK1 is activated downstream of integrin signaling pathways[76] and inactivated by binding to actopaxin, a paxillin binding protein found at focal adhesions. In response to actopaxin phosphorylation, active TESK1 is released.[77]

Antagonizing the effects of LIM and TES kinases are two specific AC phosphatases: slingshot (SSH)[78-80] and chronophin.[81] Three human SSH homologues have been identified, each expressing multiple SSH isoforms.[78,79] Each isoform exhibits unique sub-cellular localization and expression profiles,[79] allowing them to perform related yet distinct functions in various cellular and developmental processes. The SSH-1 long isoform (SSH-1L) acts as a LIMK1 phosphatase as well, thereby simultaneously activating AC and inhibiting LIMK1.[82] In contrast, p21-activated kinase 4 (PAK4) simultaneously activates LIMK1 and inactivates

SSH-1L. Much remains to be discovered concerning the mechanisms regulating other SSH isoforms, however it is certain that these complex protein kinase/phosphatase interactions and activities provide a dynamic bidirectional control mechanism governing AC phosphocycling.

In some systems, the general phosphatases PP1, PP2A and PP2B are able to dephosphorylate AC as well,[57,83] however inhibition of these phosphatases does not noticeably slow AC dephosphorylation.[57,84,85] The search for more AC phosphatases has recently unearthed chronophin phosphatase, a unique member of the haloacid dehalogenase (HAD) superfamily purified from bovine brain and found to regulate AC during the cell cycle.[81]

pH

At pH below 7.1 in vitro, cofilin[86] and ADF[27,28] do not maintain as high an actin monomer pool as at higher pH.[35] However at any pH, the ADF-actin complex has a much higher critical concentration for assembly than cofilin-actin.[35,87,88] This same behavior occurs in vivo, where it has been observed that increasing intracellular pH results in more colocalization of ADF and G-actin. In keeping with the ability of cofilin to associate with and stimulate F-actin increase in cells, the pH shift has little effect on cofilin, which remains mostly F-actin associated.[87]

Phosphatidylinositol Phosphate Binding

In vitro, phosphatidylinositol phosphates (PIP and PIP_2) inhibit the interaction between actin and many actin binding proteins including AC.[89-91] In vivo, epidermal growth factor stimulation of adenocarcinoma cells reduces PIP_2 through activation of phospholipase C, resulting in AC activation possibly by release of AC from the membrane.[34] However, an increase in intracellular calcium occurs subsequent to phospholipase C activation through the effects of the released IP_3 on calcium release channels in the endoplasmic reticulum. Calcium/calmodulin activates calcineurin, a phosphatase that can activate slingshot and thus AC by an alternative mechanism.[92]

14-3-3 Proteins

At least seven isoforms of 14-3-3 proteins occur in mammals, and these affect signaling by modulating localization, activity, or protein-protein interactions of phosphoserine-containing proteins.[93] Most phosphoserine-containing proteins will bind a number of different 14-3-3 isoforms although there is usually some isoform selectivity. Cofilin interacts with 14-3-3ζ and ε, both in vitro and in situ.[94] In addition to its phosphorylatable Ser 3, cofilin's interaction with 14-3-3 requires intact Ser 23 and Ser 24,[94] which are not phosphorylated in vivo.[18,56] Overexpression of 14-3-3ζ increases phosphorylated cofilin levels,[94] suggesting 14-3-3 protects cofilin from dephosphorylation by some phosphatases in vivo. However, 14-3-3 binding does not prevent cofilin dephosphorylation by purified SSH-1L in vitro or by expressed SSH-1L in vivo.[82]

In addition to interacting directly with cofilin, 14-3-3 also interacts with the AC kinases, LIMK1 and TESK1.[76,95,96] 14-3-3 does not stimulate LIMK phosphorylation of cofilin,[94] nor does it inhibit SSH-1L dephosphorylation of cofilin or LIMK1.[82] 14-3-3β inhibits TESK activity and its ability to be activated by integrin-mediated release from focal adhesions.[95]

Tropomyosins

Tropomyosins (TMs) are α-helical coiled-coil proteins that cooperatively bind along actin filaments. In yeast, TMs alter actin structure such that AC-dependent depolymerization and severing are inhibited. Yeast TM antagonizes the function of AC in the contractile ring.[97] TM may also block Arp2/3-mediated branching. The situation in mammals, however, is considerably more complex.[98-101]

TMs in mammals are transcribed from 4 different genes (α, β, γ and δ), which, with alternative splicing, can produce >40 different isoforms[102] that are classified into higher (~284 amino acids) and lower (247 amino acids) molecular weight groups. Stable overexpression of the nonmuscle γ-TM gene product, TM5NM1, in neuroblastoma cells results in large, spread cells with abundant contractile filaments and a diffusely staining increased inactive (phosphorylated) AC pool.[103] In these cells, transient expression of TMBr3, a much weaker F-actin binding isoform derived from the α-TM gene, decreases stress fibers and active myosin II, and promotes formation of lamellipodia containing AC and TMBr3.[103] These results suggest that some TM isoforms might cooperate with AC to turn over actin filaments.

Actin Interacting Protein 1

Actin interacting protein 1 (Aip1; also known as WD-repeat protein 1 or WDR1 in chick and mammals and unc78 in *C. elegans*) is an AC regulating protein, first identified in yeast.[104] It caps barbed ends of AC-bound F-actin, preventing filament annealing,[105] and it enhances severing activity of AC.[106,107] Several Aip1 homologues have been discovered in metazoans.[108] Chick Aip1 is rapidly upregulated in noise damaged chick cochlea and associates with ADF and actin structures,[109] as it also does in PC12 cells.[110] Aip1 supports mammalian mitotic cell rounding, and the elimination of Aip1 by siRNA impairs cell migration and cytokinesis.[111] Thus its ability to enhance AC activity is important physiologically.

Srv2/CAP

The Srv2/cyclase associated protein (CAP) family in yeast forms a high molecular weight complex linked to actin filaments via Abp1.[112] This Srv2 complex catalytically accelerates AC-dependent actin turnover in two ways: by releasing AC from ADP-actin monomers and by enhancing profilin mediated nucleotide exchange on actin monomers. Cyclase activated proteins (CAPs) are a family of highly conserved actin monomer binding proteins found in all eukaryotes.[113] Two isoforms, CAP1 and CAP2, are expressed in a cell type specific manner. Knockdown of CAP1 results in decreased actin filament depolymerization and improper AC sub-cellular localization leading to defects in cellular morphology, migration and endocytosis. Finally CAP1/ASP56, the human homologue to yeast Srv2/CAP1, was also found to enhance F-actin depolymerization and to increase the rate of G-actin nucleotide exchange.[114] Thus CAPs are important regulators of AC mediated actin filament dynamics.

Profilin

Profilin is a small actin monomer binding protein that can accelerate the rate of ADP/ATP exchange on the monomers by 1000-fold. Profilin can also deliver ATP-G-actin to filament plus ends, thus promoting actin filament assembly.[115] In the absence of free plus ends, however, profilin functions as an actin monomer sequestering protein (see ref. 116 for an extensive review of profilin). Profilin along with AC, Arp2/3 complex, Ena/VASP and capping protein were identified as the essential proteins needed to maintain *Listeria* comet-tail motility in vitro,[117] a model for membrane protrusive activity.[118] Within the cell this enhanced assembly is presumably regulated through interactions with barbed end associated assembly factors that have profilin binding domains, such as the Ena/VASP proteins or the Diaphanous related formins.[119,120]

Cortactin

Cortactin is an F-actin binding protein that localizes to sites where actin is dynamic. It is thought to assist in formation of branched filament networks by activating the Arp2/3 branching complex.[121-123] Cortactin preferentially associates with newly polymerized actin filaments (ATP-F-actin or ADP-Pi-F-actin),[124] aiding formation of branch points near extending tips of filaments and limiting the ability of AC proteins to disassemble them until cortactin dissociates. This process is of great importance in lamellipodia dynamics, as branched actin networks must cooperate to push the leading edge membrane forward as well as disassemble quickly to change the direction of motility.

Role of ADF/Cofilin Proteins in Cell Biology and Development

Cell division, cell motility, neuronal pathfinding, membrane dynamics, and cell polarization are all processes that depend heavily on actin dynamics. Much of what we know of the role of AC proteins in cells has come from studies in which deletion or mutation of AC has resulted in obvious defects in these processes. In reviewing some of these studies, we can appreciate the vital contribution AC proteins make in developing and maintaining life at all levels.

Cytokinesis

Actin filaments arrange to form a contractile ring during cell division, which ultimately constricts and separates the daughter cells. In one of the first experiments defining AC function, deletion of the cofilin gene from the yeast *S. cerevisiae* was found to be lethal because the cells failed to divide.[9,10] AC has since been localized to the cleavage furrow of dividing cells,[85,125,126] and appears to aid in assembly and maintenance of the contractile ring.[97] In a mutant *Drosophila* line expressing reduced levels of AC protein, abnormal aggregates of actin were observed at contractile ring sites in dividing spermatocytes. These aggregates remained even after completion of the cell cycle, demonstrating that AC also plays a role in disassembly of the contractile ring.[127]

Membrane Dynamics: Golgi Tubule and Vesicle Trafficking

Studies performed in cultured rat hippocampal neurons have shown that AC proteins along with their LIM kinase regulators take part in maintaining the morphology and function of the Golgi apparatus.[62,63] The neuronal Golgi is a particularly dynamic structure with 3-5 μm long tubular processes that extend and retract over time. LIMK1 and cofilin are enriched in the Golgi membranes, with cofilin present in both its active (unphosphorylated) and inactive (phosphorylated) forms. LIMK1 is targeted to the Golgi through its LIM domains and to the growth cone by its PDZ domain. Overexpression of inactive LIMK1 results in dramatic fragmentation of the Golgi, as does overexpression of a constitutively active form of cofilin (S3A mutation). Overexpression of wild type LIMK1 changes the kinetics of trafficking vesicles from the Golgi to the cell membrane, and ultimately causes growth cone collapse and axon retraction.[63] LIM kinases and cofilin thus play a critical role in axon formation through their presence in and regulation of Golgi dynamics, however neurons also rely on these proteins acting at the leading edge of growth cones, as discussed in the next section.

Cell Polarization and Motility

Both ADF and cofilin are found at the leading edge and ruffling membrane of motile cells.[42,128] A general role for the AC proteins in the establishment and maintenance of cell polarization has been unequivocally demonstrated in fibroblast cells.[129] Similar to the membranous protrusions of neurons that lead to neurite formation, fibroblasts extend cortical lamella prior to cell migration. Overexpressing LIMK1 in fibroblasts results in the inactivation of AC at the leading edge and a consequent loss of polarity. Polarity can be restored to these cells by coexpressing the constitutively active S3A-AC mutant. Not surprisingly, both LIMK1 and SSH-1L are required for maintenance of directional migration in fibroblasts.[130]

The role of AC proteins in neuronal growth cone motility and response to guidance cues has also been studied. A recent investigation into the mechanism by which the neurotrophin BDNF (brain derived neurotrophic factor) mediates growth cone filopodial extension has implicated AC as an important player in this process.[131] Introduction of the constitutively active AC-S3A mutant into cultured embryonic chick retinal neurons caused growth cone filopodia to increase in length, mimicking the effects of BDNF treatment. Further experiments showed that BDNF indeed regulates filopodial dynamics thorough a Cdc42-mediated inhibition of ROCK-dependent phosphorylation of AC.[132] Semaphorin 3A, a chemorepulsive axonal guidance molecule, causes LIM kinase mediated phosphorylation of AC,[133] which ultimately leads to collapse of the growth cone. There are now many lines of evidence defining AC proteins as essential regulators of neuronal growth cone actin dynamics (reviewed in ref. 134), and it is likely that the establishment and maintenance of neuronal cell polarization requires modulation of AC activity at nearly every stage.

AC Proteins in Apoptosis

We have seen that AC-mediated actin cytoskeleton rearrangements are essential in many processes involving changes in cell morphology. This theme seems to persist in morphological changes that occur during apoptosis as well,[135-137] although the precise mechanisms remain unclear. In some cell types, active cofilin colocalizes with actin in highly motile apoptotic membrane blebs.[138] LIMK1 also plays a role in membrane blebbing downstream of caspase-3 activation.[139] In a surprising finding, treatment of cells with agents that induce mitochondrial-dependent apoptosis (e.g., staurosporine, etoposide) causes cofilin to be translocated from the cytosol to the mitochondrial outer membrane prior to cytochrome c release and activation of the cytoplasmic caspase cascade.[140] In fact, blocking the cofilin translocation made the cells resistant to mitochondrial-dependent apoptosis. The pathway leading to cofilin translocation has yet to be resolved, however it seems that cofilin performs dual roles during apoptosis: an early role supporting cytochrome c release, and a later role in the cytosol establishing and maintaining apoptotic blebs.

Conclusions and Future Perspectives

Actin plays a central role in most dynamic cellular processes and thus it is not surprising to find important roles for AC proteins in a myriad of biological systems. However, AC proteins do not function alone in regulating actin dynamics, and it is the complex interplay of AC with actin and with the AC regulatory machinery that controls spatial and temporal changes in actin filament turnover. Furthermore, finding new functions for actin within the cell that are modulated by AC, such as Golgi membrane dynamics and mitochondrial leakage of cytochrome c, suggests that we have only begun to scratch the surface in our understanding of signaling pathways that can regulate AC proteins. This is particularly true of the pathways that regulate nuclear targeting of actin. Roles for nuclear actin are emerging as part of chromatin remodeling complexes and in transcription, and the AC proteins, with their nuclear localization signal, are prime candidates for delivery of actin to the nucleus. Whether actin remains associated with AC or dissociates from it to aid in these nuclear functions remains to be elucidated.

Acknowledgements

Special thanks to Dr. Barbara Bernstein, Dr. O'Neil Wiggan, Dr. Carin Loewen, Dr. Mike Maloney, Chi Pak and Kevin Flynn for helpful discussion and advice in preparation of this chapter. Thanks also to Dr. Allen Landes, Annette Landes and all members of the Reist and Bamburg laboratories for technical support. We gratefully acknowledge grant support from the National Institutes of Health NS48719 (to JDF) and NS40371, NS43115, DK69408 and HL58064 (to JRB).

References

1. Abe H, Obinata T. An actin-depolymerizing protein in embryonic chicken skeletal muscle: Purification and characterization. J Biochem (Tokyo) 1989; 106(1):172-180.
2. Bamburg JR, Harris HE, Weeds AG. Partial purification and characterization of an actin depolymerizing factor from brain. FEBS Lett 1980; 121(1):178-182.
3. Maekawa S, Nishida E, Ohta Y et al. Isolation of low molecular weight actin-binding proteins from porcine brain. J Biochem (Tokyo) 1984; 95(2):377-385.
4. Mabuchi I. Purification from starfish eggs of a protein that depolymerizes actin. J Biochem (Tokyo) 1981; 89(4):1341-1344.
5. Berl S, Chou M, Mytilineou C. Actin-stimulated myosin-Mg^{2+}-ATPase inhibition by brain protein. J Neurochem 1983; 40(5):1397-1405.
6. Nishida E, Maekawa S, Muneyuki E et al. Action of a 19K protein from porcine brain on actin polymerization: A new functional class of actin-binding proteins. J Biochem (Tokyo) 1984; 95(2):387-398.

7. Nishida E, Muneyuki E, Maekawa S et al. An actin-depolymerizing protein (destrin) from porcine kidney. Its action on F-actin containing or lacking tropomyosin. Biochemistry 1985; 24(23):6624-6630.
8. Cooper JA, Blum JD, Williams Jr RC et al. Purification and characterization of actophorin, a new 15,000-dalton actin-binding protein from Acanthamoeba castellanii. J Biol Chem 1986; 261(1):477-485.
9. Moon AL, Janmey PA, Louie KA et al. Cofilin is an essential component of the yeast cortical cytoskeleton. J Cell Biol 1993; 120(2):421-435.
10. Iida K, Moriyama K, Matsumoto S et al. Isolation of a yeast essential gene, COF1, that encodes a homologue of mammalian cofilin, a low-M(r) actin-binding and depolymerizing protein. Gene 1993; 124(1):115-120.
11. Hotulainen P, Paunola E, Vartiainen MK et al. Actin-depolymerizing factor and cofilin-1 play overlapping roles in promoting rapid F-actin depolymerization in mammalian nonmuscle cells. Mol Biol Cell 2005; 16(2):649-664.
12. Minamide LS, Painter WB, Schevzov G et al. Differential regulation of actin depolymerizing factor and cofilin in response to alterations in the actin monomer pool. J Biol Chem 1997; 272(13):8303-8309.
13. Gurniak CB, Perlas E, Witke W. The actin depolymerizing factor n-cofilin is essential for neural tube morphogenesis and neural crest cell migration. Dev Biol 2005; 278(1):231-241.
14. Ikeda S, Cunningham LA, Boggess D et al. Aberrant actin cytoskeleton leads to accelerated proliferation of corneal epithelial cells in mice deficient for destrin (actin depolymerizing factor). Hum Mol Genet 2003; 12(9):1029-1037.
15. Lappalainen P, Kessels MM, Cope MJ et al. The ADF homology (ADF-H) domain: A highly exploited actin-binding module. Mol Biol Cell 1998; 9(8):1951-1959.
16. Lappalainen P, Fedorov EV, Fedorov AA et al. Essential functions and actin-binding surfaces of yeast cofilin revealed by systematic mutagenesis. EMBO J 1997; 16(18):5520-5530.
17. Yonezawa N, Homma Y, Yahara I et al. A short sequence responsible for both phosphoinositide binding and actin binding activities of cofilin. J Biol Chem 1991; 266(26):17218-17221.
18. Agnew BJ, Minamide LS, Bamburg JR. Reactivation of phosphorylated actin depolymerizing factor and identification of the regulatory site. J Biol Chem 1995; 270(29):17582-17587.
19. Matsuzaki F, Matsumoto S, Yahara I et al. Cloning and characterization of porcine brain cofilin cDNA. Cofilin contains the nuclear transport signal sequence. J Biol Chem 1988; 263(23):11564-11568.
20. Iida K, Matsumoto S, Yahara I. The KKRKK sequence is involved in heat shock-induced nuclear translocation of the 18-kDa actin-binding protein, cofilin. Cell Struct Funct 1992; 17(1):39-46.
21. Abe H, Nagaoka R, Obinata T. Cytoplasmic localization and nuclear transport of cofilin in cultured myotubes. Exp Cell Res 1993; 206(1):1-10.
22. McGough A, Pope B, Chiu W et al. Cofilin changes the twist of F-actin: Implications for actin filament dynamics and cellular function. J Cell Biol 1997; 138(4):771-781.
23. Galkin VE, Orlova A, Lukoyanova N et al. Actin depolymerizing factor stabilizes an existing state of F-actin and can change the tilt of F-actin subunits. J Cell Biol 2001; 153(1):75-86.
24. Bobkov AA, Muhlrad A, Shvetsov A et al. Cofilin (ADF) affects lateral contacts in F-actin. J Mol Biol 2004; 337(1):93-104.
25. Nishida E. Opposite effects of cofilin and profilin from porcine brain on rate of exchange of actin-bound adenosine 5'-triphosphate. Biochemistry 1985; 24(5):1160-1164.
26. Carlier MF, Laurent V, Santolini J et al. Actin depolymerizing factor (ADF/cofilin) enhances the rate of filament turnover: Implication in actin-based motility. J Cell Biol 1997; 136(6):1307-1322.
27. Hawkins M, Pope B, Maciver SK et al. Human actin depolymerizing factor mediates a pH-sensitive destruction of actin filaments. Biochemistry 1993; 32(38):9985-9993.
28. Hayden SM, Miller PS, Brauweiler A et al. Analysis of the interactions of actin depolymerizing factor with G- and F-actin. Biochemistry 1993; 32(38):9994-10004.
29. Bobkov AA, Muhlrad A, Pavlov DA et al. Cooperative effects of cofilin (ADF) on actin structure suggest allosteric mechanism of cofilin function. J Mol Biol 2006; 356(2):325-334.
30. Bamburg JR, McGough A, Ono S. Putting a new twist on actin: ADF/cofilins modulate actin dynamics. Trends Cell Biol 1999; 9(9):364-370.
31. Chen H, Bernstein BW, Bamburg JR. Regulating actin-filament dynamics in vivo. Trends Biochem Sci 2000; 25(1):19-23.
32. Zebda N, Bernard O, Bailly M et al. Phosphorylation of ADF/cofilin abolishes EGF-induced actin nucleation at the leading edge and subsequent lamellipod extension. J Cell Biol 2000; 151(5):1119-1128.
33. Ghosh M, Song X, Mouneimne G et al. Cofilin promotes actin polymerization and defines the direction of cell motility. Science 2004; 304(5671):743-746.
34. Mouneimne G, Soon L, DesMarais V et al. Phospholipase C and cofilin are required for carcinoma cell directionality in response to EGF stimulation. J Cell Biol 2004; 166(5):697-708.

35. Chen H, Bernstein BW, Sneider JM et al. In vitro activity differences between proteins of the ADF/cofilin family define two distinct subgroups. Biochemistry 2004; 43(22):7127-7142.
36. Devineni N, Minamide LS, Niu M et al. A quantitative analysis of G-actin binding proteins and the G-actin pool in developing chick brain. Brain Res 1999; 823(1-2):129-140.
37. Nagaoka R, Minami N, Hayakawa K et al. Quantitative analysis of low molecular weight G-actin-binding proteins, cofilin, ADF and profilin, expressed in developing and degenerating chicken skeletal muscles. J Muscle Res Cell Motil 1996; 17(4):463-473.
38. Schuler H, Mueller AK, Matuschewski K. A Plasmodium actin-depolymerizing factor that binds exclusively to actin monomers. Mol Biol Cell 2005; 16(9):4013-4023.
39. Schuler H, Mueller AK, Matuschewski K. Unusual properties of Plasmodium falciparum actin: New insights into microfilament dynamics of apicomplexan parasites. FEBS Lett 2005; 579(3):655-660.
40. Ono S, Benian GM. Two Caenorhabditis elegans actin depolymerizing factor/cofilin proteins, encoded by the unc-60 gene, differentially regulate actin filament dynamics. J Biol Chem 1998; 273(6):3778-3783.
41. Ono S, Minami N, Abe H et al. Characterization of a novel cofilin isoform that is predominantly expressed in mammalian skeletal muscle. J Biol Chem 1994; 269(21):15280-15286.
42. Bamburg JR, Bray D. Distribution and cellular localization of actin depolymerizing factor. J Cell Biol 1987; 105(6 Pt 1):2817-2825.
43. Abe H, Ohshima S, Obinata T. A cofilin-like protein is involved in the regulation of actin assembly in developing skeletal muscle. J Biochem (Tokyo) 1989; 106(4):696-702.
44. Morgan TE, Lockerbie RO, Minamide LS et al. Isolation and characterization of a regulated form of actin depolymerizing factor. J Cell Biol 1993; 122(3):623-633.
45. Hayakawa K, Minami N, Ono S et al. Increased expression of cofilin in dystrophic chicken and mouse skeletal muscles. J Biochem (Tokyo) 1993; 114(4):582-587.
46. Shinagawa Y, Abe H, Saiga K et al. Increased expression of cofilin in denervated chicken skeletal muscle. Zool Sci 1993; 10:611-618.
47. Lopez I, Anthony RG, Maciver SK et al. Pollen specific expression of maize genes encoding actin depolymerizing factor-like proteins. Proc Natl Acad Sci USA 1996; 93(14):7415-7520.
48. Kropf DL, Bisgrove SR, Hable WE. Cytoskeletal control of polar growth in plant cells. Curr Opin Cell Biol 1998; 10(1):117-122.
49. Pendleton A, Pope B, Weeds A et al. Latrunculin B or ATP depletion induces cofilin-dependent translocation of actin into nuclei of mast cells. J Biol Chem 2003; 278(16):14394-14400.
50. Nishida E, Iida K, Yonezawa N et al. Cofilin is a component of intranuclear and cytoplasmic actin rods induced in cultured cells. Proc Natl Acad Sci USA 1987; 84(15):5262-5266.
51. Minamide LS, Striegl AM, Boyle JA et al. Neurodegenerative stimuli induce persistent ADF/cofilin-actin rods that disrupt distal neurite function. Nat Cell Biol 2000; 2(9):628-636.
51a. Bernstein BW, Chen H, Boyle JA et al. Formation of actin-ADF/cofilin rods transiently retards decline of mitochondrial potential and ATP in stressed neurons. Am J Physiol Cell Physiol 2006; [Epub ahead of print].
52. Bernstein BW, Bamburg JR. Actin-ATP hydrolysis is a major energy drain for neurons. J Neurosci 2003; 23(1):1-6.
53. Ressad F, Didry D, Xia GX et al. Kinetic analysis of the interaction of actin-depolymerizing factor (ADF)/cofilin with G- and F-actins. Comparison of plant and human ADFs and effect of phosphorylation. J Biol Chem 1998; 273(33):20894-20902.
54. Moriyama K, Iida K, Yahara I. Phosphorylation of Ser-3 of cofilin regulates its essential function on actin. Genes Cells 1996; 1(1):73-86.
55. Nebl G, Meuer SC, Samstag Y. Dephosphorylation of serine 3 regulates nuclear translocation of cofilin. J Biol Chem 1996; 271(42):26276-26280.
56. Kanamori T, Suzuki M, Titani K. Complete amino acid sequences and phosphorylation sites, determined by Edman degradation and mass spectrometry, of rat parotid destrin- and cofilin-like proteins. Arch Oral Biol 1998; 43(12):955-967.
57. Meberg PJ, Ono S, Minamide LS et al. Actin depolymerizing factor and cofilin phosphorylation dynamics: Response to signals that regulate neurite extension. Cell Motil Cytoskeleton 1998; 39(2):172-190.
58. Blanchoin L, Robinson RC, Choe S et al. Phosphorylation of Acanthamoeba actophorin (ADF/cofilin) blocks interaction with actin without a change in atomic structure. J Mol Biol 2000; 295(2):203-211.
59. Smertenko AP, Jiang CJ, Simmons NJ et al. Ser6 in the maize actin-depolymerizing factor, ZmADF3, is phosphorylated by a calcium-stimulated protein kinase and is essential for the control of functional activity. Plant J 1998; 14(2):187-193.
60. Arber S, Barbayannis FA, Hanser H et al. Regulation of actin dynamics through phosphorylation of cofilin by LIM-kinase. Nature 1998; 393(6687):805-809.

61. Yang N, Higuchi O, Ohashi K et al. Cofilin phosphorylation by LIM-kinase 1 and its role in Rac-mediated actin reorganization. Nature 1998; 393(6687):809-812.
62. Rosso S, Bollati F, Bisbal M et al. LIMK1 regulates Golgi dynamics, traffic of Golgi-derived vesicles, and process extension in primary cultured neurons. Mol Biol Cell 2004; 15(7):3433-3449.
63. Foletta VC, Moussi N, Sarmiere PD et al. LIM kinase 1, a key regulator of actin dynamics, is widely expressed in embryonic and adult tissues. Exp Cell Res 2004; 294(2):392-405.
64. Bernard O, Ganiatsas S, Kannourakis G et al. Kiz-1, a protein with LIM zinc finger and kinase domains, is expressed mainly in neurons. Cell Growth Differ 1994; 5(11):1159-1171.
65. Mizuno K, Okano I, Ohashi K et al. Identification of a human cDNA encoding a novel protein kinase with two repeats of the LIM/double zinc finger motif. Oncogene 1994; 9(6):1605-1612.
66. Acevedo K, Moussi N, Li R et al. LIM kinase 2 is widely expressed in all tissues. J Histochem Cytochem 2006; 54(5):487-501.
67. Amano T, Tanabe K, Eto T et al. LIM-kinase 2 induces formation of stress fibres, focal adhesions and membrane blebs, dependent on its activation by Rho-associated kinase-catalysed phosphorylation at threonine-505. Biochem J 2001; 354(Pt 1):149-159.
68. Edwards DC, Sanders LC, Bokoch GM et al. Activation of LIM-kinase by Pak1 couples Rac/Cdc42 GTPase signalling to actin cytoskeletal dynamics. Nat Cell Biol 1999; 1(5):253-259.
69. Nunoue K, Ohashi K, Okano I et al. LIMK-1 and LIMK-2, two members of a LIM motif-containing protein kinase family. Oncogene 1995; 11(4):701-710.
70. Sumi T, Matsumoto K, Nakamura T. Specific activation of LIM kinase 2 via phosphorylation of threonine 505 by ROCK, a Rho-dependent protein kinase. J Biol Chem 2001; 276(1):670-676.
71. Sumi T, Matsumoto K, Shibuya A et al. Activation of LIM kinases by myotonic dystrophy kinase-related Cdc42-binding kinase alpha. J Biol Chem 2001; 276(25):23092-23096.
72. Hiraoka J, Okano I, Higuchi O et al. Self-association of LIM-kinase 1 mediated by the interaction between an N-terminal LIM domain and a C-terminal kinase domain. FEBS Lett 1996; 399(1-2):117-121.
73. Edwards DC, Gill GN. Structural features of LIM kinase that control effects on the actin cytoskeleton. J Biol Chem 1999; 274(16):11352-11361.
74. Toshima J, Ohashi K, Okano I et al. Identification and characterization of a novel protein kinase, TESK1, specifically expressed in testicular germ cells. J Biol Chem 1995; 270(52):31331-31337.
75. Toshima J, Koji T, Mizuno K. Stage-specific expression of testis-specific protein kinase 1 (TESK1) in rat spermatogenic cells. Biochem Biophys Res Commun 1998; 249(1):107-112.
76. Toshima J, Toshima JY, Amano T et al. Cofilin phosphorylation by protein kinase testicular protein kinase 1 and its role in integrin-mediated actin reorganization and focal adhesion formation. Mol Biol Cell 2001; 12(4):1131-1145.
77. LaLonde DP, Brown MC, Bouverat BP et al. Actopaxin interacts with TESK1 to regulate cell spreading on fibronectin. J Biol Chem 2005; 280(22):21680-21688.
78. Niwa R, Nagata-Ohashi K, Takeichi M et al. Control of actin reorganization by Slingshot, a family of phosphatases that dephosphorylate ADF/cofilin. Cell 2002; 108(2):233-246.
79. Ohta Y, Kousaka K, Nagata-Ohashi K et al. Differential activities, subcellular distribution and tissue expression patterns of three members of Slingshot family phosphatases that dephosphorylate cofilin. Genes Cells 2003; 8(10):811-824.
80. Nagata-Ohashi K, Ohta Y, Goto K et al. A pathway of neuregulin-induced activation of cofilin-phosphatase Slingshot and cofilin in lamellipodia. J Cell Biol 2004; 165(4):465-471.
81. Gohla A, Birkenfeld J, Bokoch GM. Chronophin, a novel HAD-type serine protein phosphatase, regulates cofilin-dependent actin dynamics. Nat Cell Biol 2005; 7(1):21-29.
82. Soosairajah J, Maiti S, Wiggan O et al. Interplay between components of a novel LIM kinase-slingshot phosphatase complex regulates cofilin. EMBO J 2005; 24(3):473-486.
83. Samstag Y, Nebl G. Interaction of cofilin with the serine phosphatases PP1 and PP2A in normal and neoplastic human T lymphocytes. Adv Enzyme Regul 2003; 43:197-211.
84. Okada K, Takano-Ohmuro H, Obinata T et al. Dephosphorylation of cofilin in polymorphonuclear leukocytes derived from peripheral blood. Exp Cell Res 1996; 227(1):116-122.
85. Abe H, Obinata T, Minamide LS et al. Xenopus laevis actin-depolymerizing factor/cofilin: A phosphorylation-regulated protein essential for development. J Cell Biol 1996; 132(5):871-885.
86. Yonezawa N, Nishida E, Sakai H. pH control of actin polymerization by cofilin. J Biol Chem 1985; 260(27):14410-14412.
87. Bernstein BW, Painter WB, Chen H et al. Intracellular pH modulation of ADF/cofilin proteins. Cell Motil Cytoskeleton 2000; 47(4):319-336.
88. Yeoh S, Pope B, Mannherz HG et al. Determining the differences in actin binding by human ADF and cofilin. J Mol Biol 2002; 315(4):911-925.

89. Kusano K, Abe H, Obinata T. Detection of a sequence involved in actin-binding and phosphoinositide-binding in the N-terminal side of cofilin. Mol Cell Biochem 1999; 190(1-2):133-141.
90. Yonezawa N, Nishida E, Iida K et al. Inhibition of the interactions of cofilin, destrin, and deoxyribonuclease I with actin by phosphoinositides. J Biol Chem 1990; 265(15):8382-8386.
91. Yonezawa N, Nishida E, Iida K et al. Inhibition of actin polymerization by a synthetic dodecapeptide patterned on the sequence around the actin-binding site of cofilin. J Biol Chem 1991; 266(16):10485-10489.
92. Wang Y, Shibasaki F, Mizuno K. Calcium signal-induced cofilin dephosphorylation is mediated by Slingshot via calcineurin. J Biol Chem 2005; 280(13):12683-12689.
93. Fu H, Subramanian RR, Masters SC. 14-3-3 proteins: Structure, function, and regulation. Annu Rev Pharmacol Toxicol 2000; 40:617-647.
94. Gohla A, Bokoch GM. 14-3-3 regulates actin dynamics by stabilizing phosphorylated cofilin. Curr Biol. 2002; 12(19):1704-1710.
95. Toshima JY, Toshima J, Watanabe T et al. Binding of 14-3-3beta regulates the kinase activity and subcellular localization of testicular protein kinase 1. J Biol Chem 2001; 276(46):43471-43481.
96. Birkenfeld J, Betz H, Roth D. Identification of cofilin and LIM-domain-containing protein kinase 1 as novel interaction partners of 14-3-3 zeta. Biochem J 2003; 369(Pt 1):45-54.
97. Nakano K, Mabuchi I. Actin-depolymerizing protein Adf1 is required for formation and maintenance of the contractile ring during cytokinesis in fission yeast. Mol Biol Cell 2006; 17(4):1933-1945.
98. Bernstein BW, Bamburg JR. Tropomyosin binding to F-actin protects the F-actin from disassembly by brain actin-depolymerizing factor (ADF). Cell Motil 1982; 2(1):1-8.
99. McGough A. F-actin-binding proteins. Curr Opin Struct Biol 1998; 8(2):166-176.
100. Blanchoin L, Pollard TD, Hitchcock-DeGregori SE. Inhibition of the Arp2/3 complex-nucleated actin polymerization and branch formation by tropomyosin. Curr Biol 2001; 11(16):1300-1304.
101. Ono S, Ono K. Tropomyosin inhibits ADF/cofilin-dependent actin filament dynamics. J Cell Biol 2002; 156(6):1065-1076.
102. Gunning PW, Schevzov G, Kee AJ et al. Tropomyosin isoforms: Divining rods for actin cytoskeleton function. Trends Cell Biol 2005; 15(6):333-341.
103. Bryce NS, Schevzov G, Ferguson V et al. Specification of actin filament function and molecular composition by tropomyosin isoforms. Mol Biol Cell 2003; 14(3):1002-1016.
104. Rodal AA, Tetreault JW, Lappalainen P et al. Aip1p interacts with cofilin to disassemble actin filaments. J Cell Biol 1999; 145(6):1251-1264.
105. Okada K, Blanchoin L, Abe H et al. Xenopus actin-interacting protein 1 (XAip1) enhances cofilin fragmentation of filaments by capping filament ends. J Biol Chem 2002; 277(45):43011-43016.
106. Mohri K, Ono S. Actin filament disassembling activity of Caenorhabditis elegans actin-interacting protein 1 (UNC-78) is dependent on filament binding by a specific ADF/cofilin isoform. J Cell Sci 2003; 116(Pt 20):4107-4118.
107. Ono S, Mohri K, Ono K. Microscopic evidence that actin-interacting protein 1 actively disassembles actin-depolymerizing factor/Cofilin-bound actin filaments. J Biol Chem 2004; 279(14):14207-14212.
108. Voegtli WC, Madrona AY, Wilson DK. The structure of Aip1p, a WD repeat protein that regulates Cofilin-mediated actin depolymerization. J Biol Chem 2003; 278(36):34373-34379.
109. Oh SH, Adler HJ, Raphael Y et al. WDR1 colocalizes with ADF and actin in the normal and noise-damaged chick cochlea. J Comp Neurol 2002; 448(4):399-409.
110. Shin DH, Lee E, Chung YH et al. Subcellular localization of WD40 repeat 1 protein in PC12 rat pheochromocytoma cells. Neurosci Lett 2004; 367(3):399-403.
111. Fujibuchi T, Abe Y, Takeuchi T et al. AIP1/WDR1 supports mitotic cell rounding. Biochem Biophys Res Commun 2005; 327(1):268-275.
112. Balcer HI, Goodman AL, Rodal AA et al. Coordinated regulation of actin filament turnover by a high-molecular-weight Srv2/CAP complex, cofilin, profilin, and Aip1. Curr Biol 2003; 13(24):2159-2169.
113. Bertling E, Hotulainen P, Mattila PK et al. Cyclase-associated protein 1 (CAP1) promotes cofilin-induced actin dynamics in mammalian nonmuscle cells. Mol Biol Cell 2004; 15(5):2324-2334.
114. Moriyama K, Yahara I. Human CAP1 is a key factor in the recycling of cofilin and actin for rapid actin turnover. J Cell Sci 2002; 115(Pt 8):1591-1601.
115. Gutsche-Perelroizen I, Lepault J, Ott A et al. Filament assembly from profilin-actin. J Biol Chem 1999; 274(10):6234-6243.
116. Witke W. The role of profilin complexes in cell motility and other cellular processes. Trends Cell Biol 2004; 14(8):461-469.
117. Loisel TP, Boujemaa R, Pantaloni D et al. Reconstitution of actin-based motility of Listeria and Shigella using pure proteins. Nature 1999; 401(6753):613-316.
118. Carlier MF, Wiesner S, Le Clainche C et al. Actin-based motility as a self-organized system: Mechanism and reconstitution in vitro. C R Biol 2003; 326(2):161-170.

119. Laurent V, Loisel TP, Harbeck B et al. Role of proteins of the Ena/VASP family in actin-based motility of Listeria monocytogenes. J Cell Biol 1999; 144(6):1245-1258.
120. Romero S, Le Clainche C, Didry D et al. Formin is a processive motor that requires profilin to accelerate actin assembly and associated ATP hydrolysis. Cell 2004; 119(3):419-429.
121. Weed SA, Karginov AV, Schafer DA et al. Cortactin localization to sites of actin assembly in lamellipodia requires interactions with F-actin and the Arp2/3 complex. J Cell Biol 2000; 151(1):29-40.
122. Weaver AM, Karginov AV, Kinley AW et al. Cortactin promotes and stabilizes Arp2/3-induced actin filament network formation. Curr Biol 2001; 11(5):370-374.
123. Uruno T, Liu J, Zhang P et al. Activation of Arp2/3 complex-mediated actin polymerization by cortactin. Nat Cell Biol 2001; 3(3):259-266.
124. Bryce NS, Clark ES, Leysath JL et al. Cortactin promotes cell motility by enhancing lamellipodial persistence. Curr Biol 2005; 15(14):1276-1285.
125. Nagaoka R, Abe H, Kusano K et al. Concentration of cofilin, a small actin-binding protein, at the cleavage furrow during cytokinesis. Cell Motil Cytoskeleton 1995; 30(1):1-7.
126. Ono K, Parast M, Alberico C et al. Specific requirement for two ADF/cofilin isoforms in distinct actin-dependent processes in Caenorhabditis elegans. J Cell Sci 2003; 116(Pt 10):2073-2085.
127. Gunsalus KC, Bonaccorsi S, Williams E et al. Mutations in twinstar, a Drosophila gene encoding a cofilin/ADF homologue, result in defects in centrosome migration and cytokinesis. J Cell Biol 1995; 131(5):1243-1259.
128. Yonezawa N, Nishida E, Koyasu S et al. Distribution among tissues and intracellular localization of cofilin, a 21kDa actin-binding protein. Cell Struct Funct 1987; 12(5):443-452.
129. Dawe HR, Minamide LS, Bamburg JR et al. ADF/Cofilin controls cell polarity during fibroblast migration. Curr Biol 2003; 13(3):252-257.
130. Nishita M, Tomizawa C, Yamamoto M et al. Spatial and temporal regulation of cofilin activity by LIM kinase and Slingshot is critical for directional cell migration. J Cell Biol 2005; 171(2):349-359.
131. Gehler S, Shaw AE, Sarmiere PD et al. Brain-derived neurotrophic factor regulation of retinal growth cone filopodial dynamics is mediated through actin depolymerizing factor/cofilin. J Neurosci 2004; 24(47):10741-10749.
132. Chen TJ, Gehler S, Shaw AE et al. Cdc42 participates in the regulation of ADF/cofilin and retinal growth cone filopodia by brain derived neurotrophic factor. J Neurobiol 2006; 66(2):103-114.
133. Aizawa H, Wakatsuki S, Ishii A et al. Phosphorylation of cofilin by LIM-kinase is necessary for semaphorin 3A-induced growth cone collapse. Nat Neurosci 2001; 4(4):367-373.
134. Sarmiere PD, Bamburg JR. Regulation of the neuronal actin cytoskeleton by ADF/cofilin. J Neurobiol 2004; 58(1):103-117.
135. Cotter TG, Lennon SV, Glynn JM et al. Microfilament-disrupting agents prevent the formation of apoptotic bodies in tumor cells undergoing apoptosis. Cancer Res 1992; 52(4):997-1005.
136. Laster SM, Mackenzie Jr JM. Bleb formation and F-actin distribution during mitosis and tumor necrosis factor-induced apoptosis. Microsc Res Tech 1996; 34(3):272-280.
137. Rao JY, Jin YS, Zheng Q et al. Alterations of the actin polymerization status as an apoptotic morphological effector in HL-60 cells. J Cell Biochem 1999; 75(4):686-697.
138. Mannherz HG, Gonsior SM, Gremm D et al. Activated cofilin colocalises with Arp2/3 complex in apoptotic blebs during programmed cell death. Eur J Cell Biol 2005; 84(4):503-515.
139. Tomiyoshi G, Horita Y, Nishita M et al. Caspase-mediated cleavage and activation of LIM-kinase 1 and its role in apoptotic membrane blebbing. Genes Cells 2004; 9(6):591-600.
140. Chua BT, Volbracht C, Tan KO et al. Mitochondrial translocation of cofilin is an early step in apoptosis induction. Nat Cell Biol 2003; 5(12):1083-1089.
141. Maciver SK, Hussey PJ. The ADF/cofilin family: Actin-remodeling proteins. Genome Biol 2002; 3(5):3007.1-3007.12, (reviews).
142. Jang DH, Han JH, Lee SH et al. Cofilin expression induces cofilin-actin rod formation and disrupts synaptic structure and function in Aplysia synapses. Proc Natl Acad Sci USA 2005; 102(44):16072-16077.
143. Aizawa H, Kishi Y, Iida K et al. Cofilin-2, a novel type of cofilin, is expressed specifically at aggregation stage of Dictyostelium discoideum development. Genes Cells 2001; 6(10):913-921.
144. Bowman GD, Nodelman IM, Hong Y et al. A comparative structural analysis of the ADF/cofilin family. Proteins 2000; 41(3):374-384.
145. Hatanaka H, Ogura K, Moriyama K et al. Tertiary structure of destrin and structural similarity between two actin-regulating protein families. Cell 1996; 85(7):1047-1055.
146. Fedorov AA, Lappalainen P, Fedorov EV et al. Structure determination of yeast cofilin. Nat Struct Biol 1997; 4(5):366-369.
147. Pope BJ, Zierler-Gould KM, Kuhne R et al. Solution structure of human cofilin: Actin binding, pH sensitivity, and relationship to actin-depolymerizing factor. J Biol Chem 2004; 279(6):4840-4848.

Chapter 3

Profilin, an Essential Control Element for Actin Polymerization

Roger Karlsson* and Uno Lindberg

Abstract

This chapter reviews some aspects of the biochemistry and cellular function of profilin, focussing on its role as a control component of actin polymerization. Signalling-dependent changes in cell behaviour are direct consequences of a force-generating remodelling of the actin microfilament system at the inner surface of the plasma membrane. Characteristic for this sub-membraneous region is the enrichment of actin filaments in highly ordered bundles and sheets of filaments. These filaments, which have their fast polymerizing (+)-ends facing the lipid bilayer, are under constant turnover, with ATP-containing actin monomers being added at their (+)-ends, and ADP-actin monomers dissociating from the (-)-ends in a treadmilling process regulated by a number of actin-binding proteins. Here, profilin comes into play as one of the key regulators of actin filament formation. The protrusive surface activity, typically seen after receptor stimulation, is caused by local incorporation of actin from profilin-actin into the ends of growing filaments of filopodia and lamellipodia. Thus, the function of profilin is primarily integrated with the force-generating microfilament apparatus at the cell periphery. In addition to actin, profilin also binds polyphosphoinositides and proline-rich ligands.

Profilin and Profilin:Actin

The identification of the DNase I inhibitor from bovine spleen as nonfilamentous actin and subsequent analysis of crystals of the inhibitor revealed the presence of a protein other than actin in the crystals.[1-4] This protein turned out to be an efficient inhibitor of actin polymerization under certain conditions. It was therefore called profilin, since it seemed obvious that its function would be to keep the actin in an unpolymerized, profilamentous form.[4,5] The name profilactin for the complex was adopted with reference to an unpolymerized form of actin found by Tilney in the subacrosomal cup of *Thyone* sperm.[6] The subacrosomal cup consists of actin, two high molecular weight spectrin-like proteins and a small molecular weight component originally thought to be contaminating protamines.[6,7] This protein was later proven to be a *Thyone* profilin.[8] Thus, *Thyone* sperm provides one of the clearest examples of an in vivo situation, where profilin:actin undoubtedly deliveres actin for filament formation. Subsequently, profilin from Acanthamoeba was isolated, and it was demonstrated that profilin inhibits nucleation of filament formation in vitro, but that filament elongation is less efficiently interfered with.[9,10]

Evidence for a filament precursor-role of profilin:actin was also obtained from studies of platelet activation.[11] Actin filaments are asymmetric having a fast and a slow polymerizing end, here referred to as the (+)- and (-)-ends.[12] Experiments using preformed nuclei, with either the

*Corresponding Author: Roger Karlsson—Department of Cell Biology, Stockholm University, S-106 91, Stockholm, Sweden. Email: roger.karlsson@cellbio.su.se

Actin-Monomer-Binding Proteins, edited by Pekka Lappalainen. ©2007 Landes Bioscience and Springer Science+Business Media.

(+)-or the (-)-end blocked, demonstrated that profilin only efficiently inhibits elongation at the (-)-end.[8,13-15] The explanation to this behaviour came from solving the structure of the profilin I:β-actin complex, which revealed that profilin binds to the end of the actin monomer corresponding to that exposed at the filament (+)-end, leaving the other end of the monomer free to bind to the (+)-end of the growing filament.[16,17] In combination with (+)-end capping proteins, no interaction between profilin:actin and filaments can occur and profilin then works as an actin sequestering protein, promoting depolymerization and inhibition of the formation of actin nuclei.

Isoform Diversity

Profilin is an essential protein in all eukaryotes,[18-20] and several isoforms of the protein are present in both plants and animals, for (see refs. 21,22). In mammals, five isoforms are known: profilin I, IIa, IIb, III and IV. They all bind to actin, polyphosphoinositides (PIP2 and PIP3) and proline-rich sequences, but the binding constants for these interactions vary between different isoforms. Profilin I is ubiquitous, while the tissue distribution of the others is more restricted, reflecting that different cells require profilins having somewhat different ligand-binding properties. Profilin II primarily functions in the central nervous system, but mRNA encoding this isoform has been detected in kidney and muscle,[20,23,24] and both profilin I and II can be isolated from smooth muscle (Grenklo and Karlsson, unpublished). It was shown in mice that the profilin IIa splice product is most abundantly expressed during embryogenesis at stages of rapid brain development,[25,26] and mice deficient for this isoform develop neurological defects.[20] The more rarely expressed IIb-isoform, like the profilin encoded by *Vaccinia* virus,[27] is unique among the profilins in that it has a low affinity for poly-(L-proline). Interestingly, this particular profilin is also a weak binding partner of actin, but appears to bind one or more tubulin associated proteins.[25]

Two additional splice variants of the mouse profilin II mRNA, which have the capacity to encode 55 and 61 residue long truncated forms, respectively, have been reported,[26] but whether the corresponding proteins are expressed is unknown. The distribution of the profilin isoforms III and IV, finally, appears to be restricted to testis, where there is evidence that the latter functions during spermiogenesis.[28,29]

Plant profilins constitute major pollen and food allergens.[30,31] and other profilins, with more divergent amino acid sequences are expressed by apicomplexan parasites like *Toxoplasma gondii* and *Plasmodium falciparum*.[32] Compared to mammalian profilin I, these have several sequence insertions some of which are relatively long, making these profilins the largest known so far with over 170 residues compared to 139 for human profilin I. The profilin from *Plasmodium falciparum*, expressed in *E.coli*, binds actin, although with low affinity (Herwig Schüler, personal communication).

The Profilin Structure

The presence of profilin in all eukaryotes suggests that it plays evolutionary conserved functions. In support of this, birch profilin expressed in BHK-cells distributes like the endogenous profilin and *Dictyostelium* profilin null mutants can be rescued with a plant profilin.[33,34] The profilin fold is conserved not only in different isoforms within the same organism, but also in profilins from phylogenetically distant species.[16,17,35-38] The sequence homologies are rather low even in isoforms from the same organism, but the character of the amino acid residues and their positioning in the actin-binding site are preserved.[22,37,38]

Profilin consists of a seven-stranded anti-parallel β-pleated sheet with the α-helical N- and C-termini packed on one side and two short α-helices on the other,[16,17] (Fig. 1). The mammalian profilins have a protruding loop connecting strands 5 and 6 (K90-T97), whereas in nonvertebrate profilins this loop is less prominent.[17,36,38] The loop is positioned close to the actin-binding site and deletion of two residues (P96, T97) at its base drastically lowers the affinity for actin.[39,40] Further analysis of the role of this structure for profilin function

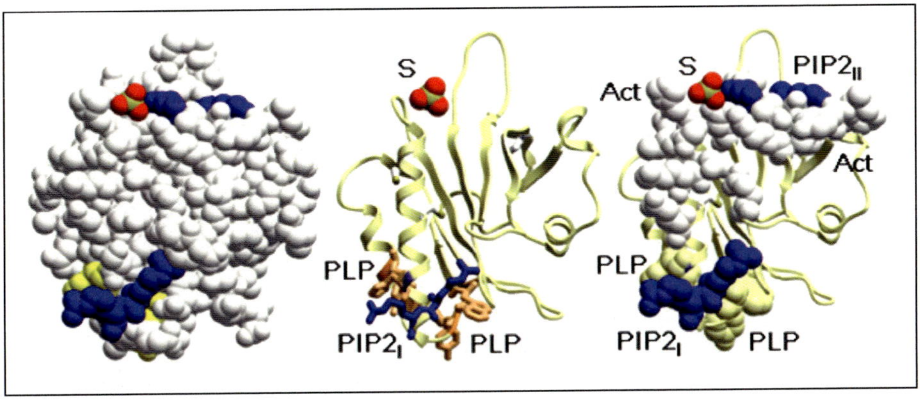

Figure 1. The profilin molecule. Left panel illustrates the interaction-surfaces for poly-L-proline (yellow) and PIP2 (blue). The centre panel highlights the positioning to the N- and C-terminal helices of the residues involved in the initial PIP2-binding (blue; PIP2$_I$) as well as the poly-L-proline-binding residues (brown; PLP). To the right, the actin-binding site (white; Act) is shown together with the site for poly-(L-proline) and the two PIP2-binding surfaces in space filling mode (light yellow and blue, respectively). The sulphate (S) in the second PIP2 site (PIP2$_{II}$) is also illustrated (red and yellow). See text for references and details. Illustrations were made in Molsoft ICM Browser.

should be of interest. The principal features of the profilin fold are also found in the endosomal adaptor protein p14 and in members of a snare subfamily, the longins, which indicates that these proteins may express some of the functions of profilin.[41,42]

Poly-(L-Proline) and Profilin

The discovery that profilin binds poly-(L-proline) was used to develop a fast procedure for the isolation of profilin and profilin:actin based on affinity chromatography on poly-(L-proline).[43,44] The first profilin-binding partner found, apart from actin, was the vasodilator-stimulated phosphoprotein (VASP), whose interaction with profilin is mediated via a proline-rich sequence-motif.[45] This binding activity is now recognized as an important feature of profilin to establish contact with a number of different protein ligands, exposing a variety of proline-rich sequence-motifs, which do not necessarily consist of only prolines. It is also noteworthy that the binding of poly-(L-proline) by profilin might occur in either polypeptide backbone orientation.[46,47] For a listing of proline-rich profilin ligands (see ref. 22).

The poly-(L-proline) binding site is formed by residues in the N- and C-terminal α-helices and overlaps partly with surfaces on the molecule engaged in binding phosphoinositide lipids.[48-50] The stability of the packing of the two terminal helices in profilin I is in part contributed for by a cluster of aromatic residues comprised of the side chains of W3, Y6, W31, Y139, and H133, (Fig. 1). These residues constitute the poly-(L-proline) binding site.[17,46,47]

The function of profilin is sensitive to modifications of the terminal helices and interference with these side chain-interactions influences its capacity to interact with its partners. Removal of the two C-terminal residues by limited proteolysis alters the solubility and lipid binding properties of profilin I,[51] nitration of Y139 increases the affinity for poly-(L-proline) and interferes with actin binding,[52] and introduction of mutations in this region either interferes with or strengthens, the poly-(L-proline) interaction.[48,53,54] Binding of phosphatidylinositol (4,5)-bisphosphate (PIP2), to this region quenches the fluorescence resonance from the two tryptophanes (W3, W31) suggesting that they move relative to each other.[55] This structural alteration causes the dissociation of the profilin:actin complex.[56] The binding of antibodies directed against the tryptophanes in *Achantamoeba* profilin, in contrast to PIP2, strengthens the actin-interaction,[57] and the subtle changes in the binding to actin seen after introduction of point mutations in the poly-(L-proline) binding surface [53] further support that there is a communication between this region and the actin binding site, presumably via the C-terminal helix.[50]

Interactions between profilin and proline-containing ligands may be regulated by phosphorylation.[53,58-61] A kinase activity, which coimmunoprecipitates with the epidermal growth factor (EGF)-receptor-complex from EGF-stimulated cells leads to mutually exclusive phosphorylations of S137 or Y139 in profilin I and abolishes its binding to poly-L-proline.[53,54] Plant profilin has been shown to be a substrate for a MAP-kinase in vitro,[62] and to become tyrosine phosphorylated in vivo, [63] It is unclear how these modifications affect the functions of the profilins. Together these observations point to the polyproline-binding region as a 'hot-spot' for signal-induced regulation of profilin functions.

Structure of Profilin: Actin and Its Role in Filament Formation

The structures of actin orthologs, cocrystallized with different actin binding proteins, all have a closed nucleotide binding cleft, and the conformation of actin appears relatively unchanged regardless of whether the nucleotide is ATP or ADP, or whether the tightly bound cation is Ca^{2+} or Mg^{2+}.[16,64-66] This might seem contradictory in view of the many observations indicating that actin can undergo conformational changes involving the nucleotide binding cleft in vitro. An explanation to the closed, rather invariable conformation could be that it is imposed on the actin by the nature of the crystal packing or ligand binding.[67,68] However, a more appealing explanation is that the common conformation of actin in solution is the closed state, and that the extreme structural variability (closed-to-open transition) seen in the case of profilin:actin reflects the physiological function of profilin, namely to

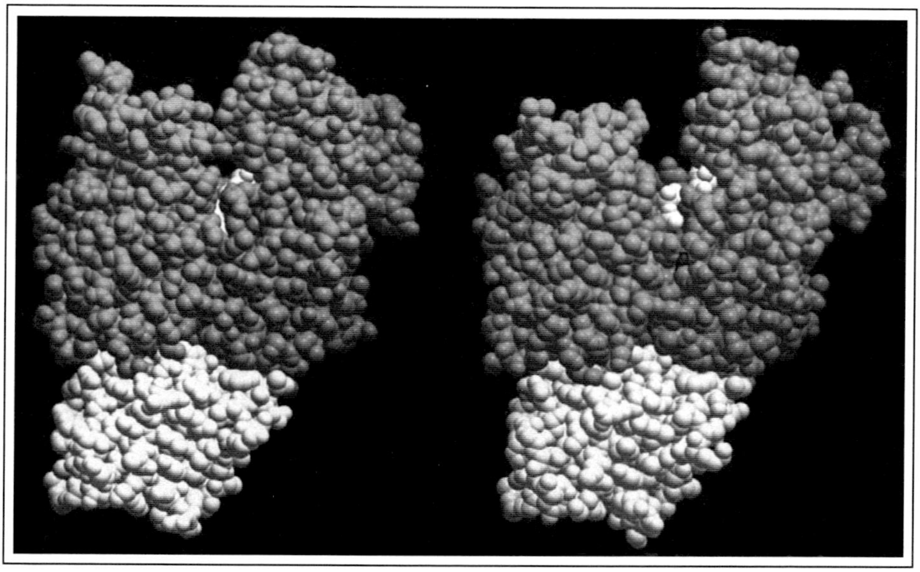

Figure 2. The profilin:actin closed and open states. The crystal structures of profilin:actin solved in 3.5 M NH_4SO_4 (closed state, left) and 1.8 M potassium phosphate (open state, right) in the presence of Ca^{2+} and ATP, illustrating the conformational difference between the two states. Actin is shown in blue and profilin in yellow; the cleft-opening (right) exposes the polyphosphate tail of the ATP (white).[67] A color version of this figure is available online at www.Eurekah.com.

open the nucleotide binding cleft, allowing nucleotide exchange (Fig. 2). Profilin in vitro efficiently accelerates nucleotide exchange on actin.[69] Even with mutants of profilin, whose interaction with actin is attenuated, there is a fast exchange of the nucleotide.[40] Since profilin also inhibits the actin ATPase activity,[10,70] this favors the delivery of ATP-actin to sites of polymerization as reviewed recently.[71]

The observation that a cross-linked form of profilin:actin, here called PxA,[70,72] in polymerization experiments interferes with the addition of actin at the (+)-end of actin filaments, without itself being incorporated in the growing filament, lends credence to the view that profilin:actin initially binds as a complex to the filament (+)-end with the actin in the open state and ATP bound to it. Once added, the profilin is released from the actin at the growing end, allowing the newly added actin to become stably incorporated into the filament. This step, thought to be connected to ATP hydrolysis and a subsequent conformational change on the actin, would entail the transition to the tight state.[71] Further evidence for the central role of profilin:actin in microfilament formation in vitro and in vivo was obtained with the use of a mutant profilin (H119E), which does not bind actin. The mutant profilin blocked WASP-dependent polymerization in vitro, and formation of filopodia in cells.[73,74] It is noteworthy that the tumor suppressor activity of profilin[75,76] is dependent on a functional actin binding site.[77]

Polyphosphoinositides and Profilin

In the original experiments showing that phosphatidylinositol-(4,5)-bisphosphate (PIP2) can dissociate profilin:actin, there was also evidence for small amounts of actin, remaining in complex with profilin on lipid micelles.[56] Recently it was demonstrated that PIP2 can bind to cross-linked profilin:actin, proving the existence of a binding site at a distance from the actin binding site.[50] Many laboratories have contributed to the characterization of the phosphoinositide-binding to profilin, see for instance.[49,78-82] The conclusion is that PIP2 can bind to two sites on profilin, one adjacent to the binding-site for poly-(L-proline) and a second which is located within the actin-binding site (Fig. 1). Dissociation of profilin:actin by PIP2 may result from a successive destabilization of the complex initiated by PIP2 binding to the positively charged residues in the C-terminal helix, perturbing the poly-(L-proline) binding region with repercussions for the actin binding.[50] Analysis of crystal structures of human profilins have identified a high affinity binding pocket for phosphate and sulphate in the second PIP2-binding site formed by R88, N99, H119 and the main chain amide of residue 120,[37] which could become accessible for binding of profilin to the negatively charged polyphosphoinositides after release of the actin (Fig. 1).

Much less is known about the role of the profilin-PIP2 interaction in vivo, and its connection to remodelling of the microfilament system in response to signal transduction. In thrombin-stimulated platelets, a rapid increase in PIP2 concentration coincides with extensive polymerization of actin from profilin:actin,[11,83-85] a process which is sensitive to drugs interfering with phosphatidylinositol metabolism.[86]

Furthermore, profilin might regulate the metabolism of the phosphoinositides via pathways connected to receptor signalling. It has been reported that profilin inhibits phospholipase Cγ (PLCγ) hydrolysis of PIP2 to inositol-trisphosphate and diacylglycerol unless the enzyme is activated by tyrosine phosphorylation.[87,88] Another connection is the binding of profilin to the regulatory subunit p85 of phosphatidylinositol 3-kinase, which increases the V_{max} of the lipid-kinase activity expressed by the catalytic p110 subunit of the enzyme,[89] leading to formation of phosphatidylinositol-(3,4,5)-trisphosphate. This inositide activates a guanidine exchange factor called Vav that operates on the small GTPase Rac, leading to increased synthesis of polyphosphoinositides, further activating the microfilament system.[90-92] It is not unlikely, that profilin, released from actin might associate with clusters of polyphosphoinositides in the inner leaflet of the plasma membrane.[93]

Profilin in Cell Motility

Fluorescence microscopy of tissue cultured cells using antibodies to profilin demonstrated that the protein is distributed all over the cell in a fine-granular pattern, with accumulations of fluorescence at the cell periphery, in the peri-nuclear area and in the nucleus.[94-97]

The accumulation of profilin at the cell edge in highly motile regions reflects its role in the control of actin polymerization, delivering actin in the form of profilin:actin to different polymer-forming protein machineries (see below). This generates the surface protrusions cells use to establish new contacts with the extracellular matrix for migration and with neighbouring cells. It has been argued that the heptameric protein complex Arp2/3, plays a pivotal role in this context as a filament-associated de novo nucleator of actin filaments.[98] This complex was first isolated from an *Acanthamoeba* extract by affinity chromatography on immobilized profilin,[99] and one of the actin-related proteins in the complex, Arp2, was identified as an interaction partner of profilin.[100] The actin nucleating activity of the complex in vitro is strongly stimulated in the presence of the WASP/Wave/Scar family of proteins and by the phosphotyrosine-regulated protein cortactin, which accumulates at the leading cell edge after receptor stimulation.[101-104] Furthermore, the Arp2/3 complex is required for the movement of the intracellular bacterium *Listeria monocytogenes*, whose surface protein ActA, like WASP/Wave/Scar,[105] activates Arp2/3 to nucleate formation of actin filaments.[106] This process drives the bacterium through the infected cytoplasm by sustained actin polymerization through the action of VASP and its recruitment of profilin:actin.[107-109]

In vitro, Arp2/3 can bind to the side of actin filaments and nucleate polymerization of new filaments, which elongate at an angle of approximately 70° with respect to the mother filament, e.g.[110-115] Based on the crystal structure of the complex, it was suggested that Arp2 and 3 within the complex reorient to form the nucleating surface.[116] Results from electron microscopy of detergent-extracted cultured cells seem to suggest the existence of branched actin filaments in lamellipodia.[117,118] This led to the dendritic branch model of actin-driven advancement of lamellipodia of cells.[98,110,119] However, there are observations which argue against this model in its present form. Analysis of cells prepared to optimize the preservation of the arrangements of actin filaments in lamellipodia revealed the presence of a dense organization of unbranched filaments, several micrometers long.[120,121] Collections of small numbers of filaments appear to converge at special structures at the very edge; not to diverge towards the membrane like in the dendritic branch model.[120-123] The density of filaments in the lamellipodium precludes the direct observation of the presence of branched filaments in the edge-zone, but in the rest of the lamellipodium, the filaments do not seem to be branched (Fig. 3). It is plausible that branches of oligomeric actin in the edge-zone are rapidly transformed into the long filaments dominating the scene in the rest of the lamellipodium. Debranching during Arp2/3-dependent polymerization of actin has been observed in vitro, but this process is rather slow, suggesting that if it occurs, there ought to be accelerating factors in vivo.[124]

There is no question that Arp2/3 has the capacity to form branched filaments in vitro, and it has been suggested that cofilin with its severing activity enhances actin polymerization in cooperation with Arp2/3 by increasing the number of actin filament (+)-ends,[125] but how these proteins cooperate in the productive formation of lamellipodia and filopodia during cell migration is still unclear. It has been demonstrated that tropomyosin inhibits branch formation, and blocks severing by cofilin in vitro.[126-128] Recently, it was reported that tropomyosin is present all the way to the tip of lamellipodia, which would interfere with branch formation in vivo.[129] Furthermore, drastically reducing the cellular concentration of Arp2/3 has very little effect on leading edge organization and cell advancement.[130] Thus, either this actin polymerizing machinery plays a minor role in forming cell surface protrusions, or other mechanisms effectively substitute for its activity. Clearly, the current dendritic branch model needs to be reevaluated.

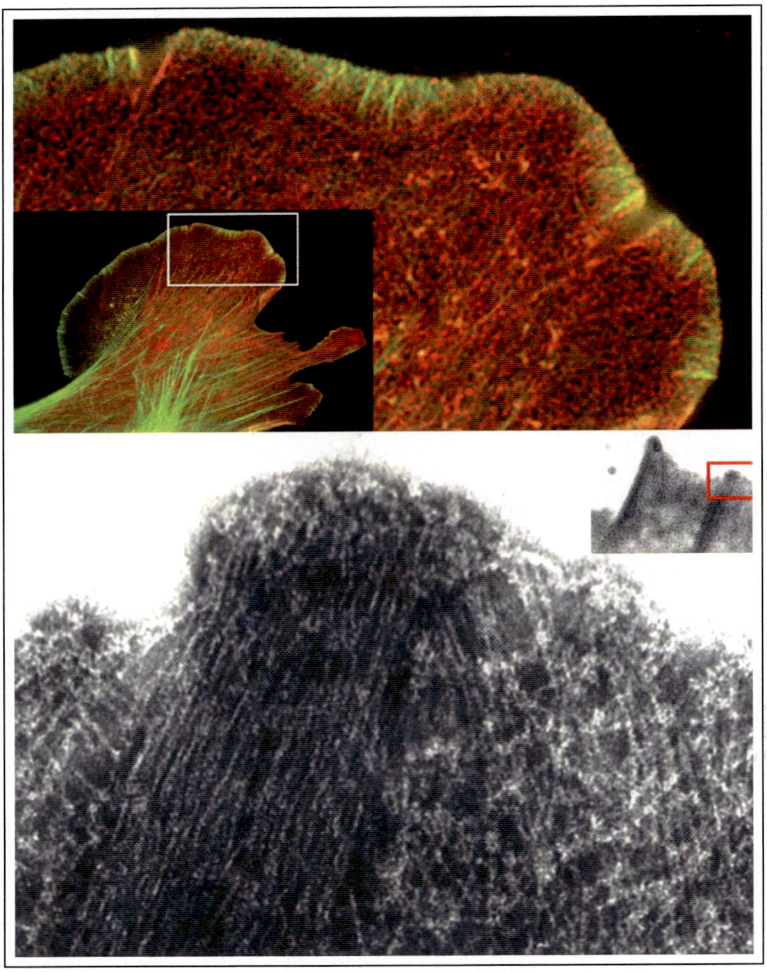

Figure 3. The organization of the microfilament system at the leading cell edge. Upper panel, shows the distribution of profilin (red) visualized by affinity purified profilin antibodies generated against PxA.[138] Filamentous actin is labelled green (FITC-phalloidin). Lower panel, a corresponding section of a cell edge seen by electron microscopy after sample preparation as described in reference 120.

Observations of the movement in cell extracts of microscopic beads functionalized with WASP suggest that filament elongation resulting from the concerted action of WASP-Arp2/3 is likely to involve components such as VASP and WIP,[131] which through profilin recruits profilin:actin for incorporation of actin into growing filaments (see below). In support of this, movement of *Listeria* in an in vitro reconstituted system is enhanced by the presence of profilin and VASP,[109] and the concentration of VASP at the cell edge appears to correlate directly with the rate by which the edge advances.[132]

The second profilin:actin-dependent actin polymerization mechanism is represented by the formin family of proteins.[133] One of the most potent actin polymerizing members of this family is mDia. A fragment of mDia containing the FH1 and FH2 domains polymerizes actin, with an efficiency greatly enhanced by profilin, leading to formation of long filaments in parallel arrays, which rapidly elongate.[134-136]

Further evidence for the precursor role of profilin:actin in actin filament formation in cells has come from the use of the nondissociable profilin:actin, PxA. Microinjection of PxA into spreading cells followed by stimulation with platelet derived growth factor (PDGF), or PxA injection into cells infected with the intracellular pathogen *Listeria monocytogenes*, interfered with actin polymerization.[72,108] This was seen as abrogated lamellipodia formation and blocked PDGF-induced ruffling, and in the case of *Listeria* as an abrupt retardation of the movement followed by dissociation and depolymerization of its characteristic 'comet tail' of actin filaments. In in vitro experiments, PxA binds weakly to actin filament (+) ends and interferes with polymerization without changing the final steady state level of filamentous actin,[70] suggesting that profilin must be released before the incoming actin can be stably incorporated into the actin helix. The pronounced effect on actin filament formation in vivo suggests that PxA blocks the actin-polymerizing machinery by binding to one of its components, presumably VASP, which is required for efficient *Listeria* motility.[107,137] The failure to release profilin, due to the covalent linkage, prohibits stable incorporation of actin into the growing filaments, causing filament breakage. The recognition of profilin:actin as the functional component by the polymer-forming machinery was further tested by making PxA with the H133S mutant profilin,[108] which is unable to recognize proline-rich sequences.[48,53] This modified PxA had no effect on the bacterial movement, nor did it affect the motile activity of the infected cells. It is therefore reasonable to believe that VASP, as part of the actin polymerizing system on the bacterial surface, would be the factor recruiting profilin:actin for filament formation. Dickinson and Purich have proposed a model, in which *Listeria* motility depends on a 'molecular clamp' that controls actin assembly at the interface between bacteria and actin tails.[139,140] We argue that PxA poisons the 'molecular clamp' causing actin-tail detachment, as described above.

Clearly, the control of actin polymerization in vivo is characterized by a high degree of complexity, incorporating elaborate mechanisms linked to transmembrane signalling, members of the Rho-family of GTPases, a number of additional proteins and possibly polyphosphoinositides, (e.g., see refs. 141-144). These activities must be coordinated with the activity of proteins controlling the availability of the filament (+)-end.[145] Proteins involved in this context are exemplified by capping protein, gelsolin, Eps8, carmil, twinfilin and tropomyosin.[146-151] As mentioned above, dissociation of profilin and final integration of new actin monomers into the filaments is likely to be coupled to hydrolysis of the actin-bound ATP. Therefore, actin filaments formed in vivo should consist of ADP-carrying actin subunits along their entire length, even under rapid elongation. The newly formed filaments have to be stabilized, for instance by the binding of tropomyosin, like in budding yeast.[152] In addition, cofilin-ADP-actin, which comes off the (-)-end of depolymerizing filaments, is rapidly recycled into profilin:ATP-actin, possibly via the action of Srv2/CAP, (e.g., see ref. 153).

There is also a question of transport of profilin:actin to sites of actin polymerization from the inner end of lamellipodia to the tip of the advancing cell edge. Diffusion would not be sufficient.[154] Instead, it must involve active transport, possibly by an unconventional myosin, as reported for localization of integrins to the tips of filopodia.[155,156] Staining of cells using antibodies generated against PxA gives rise to fluorescent dots (Fig. 3), which could represent a transport-form of profilin:actin being brought to filament-forming sites.[138] Thus, in a sense, the advancement of the cell edge is driven by ATP-hydrolysis on myosin motors transporting profilin:actin and energy-dependent insertion of actin at filament (+)-ends.

Profilin in the Nucleus

Early stainings of cells using profilin antibodies suggested the presence of profilin in the nucleus.[94,95] Now it is known that both profilin I and II distributes to the nucleus,[96,97] and apparently interacts with the protein SMN (survival of motor neuron protein),[96] which is essential for the formation of small nuclear ribonucleoproteins (sn-RNPs). Mutational inactivation of the gene encoding SMN leads to degeneration of motor neurons and spinal muscular atrophy.[157,158] It is unclear whether this is an effect connected to the interaction between profilin and SMN in the nucleus or in the nerve cell growth cone, where SMN is also found, or if the effect is unrelated to profilin. The observation that nuclear profilin I accumulates in Cajal bodies and nuclear speckles, and that profilin antibodies interferes with pre-mRNA splicing in vitro,[97] indicates that it operates in the nucleus, perhaps in pre-mRNA-splicing in connection to SMN and sn-RNP. It is noteworthy that phosphatidylinositol-4-phosphate-5-kinase is present in nuclear speckles,[159] where the enzyme may produce PIP2, which seems to be necessary for pre-mRNA splicing.[160]

Actin is found in the nucleus as well, where it seems to function in chromatin remodelling and transcription,[161,162] making it possible that also the profilin:actin complex operates in the nucleus. In support of this contention, an export protein specifically recognizing profilin:actin has been identified.[163] In addition, profilin interacts with the transcription factor c-Myb, coupling profilin to regulation of transcription.[164] Many other microfilament associated proteins have been found in the nucleus and the elucidation of the function of actin-based chemo-mechanical transduction in connection to nuclear processes is of great interest.[161]

Profilin in the Brain

Profilin has been shown to be important for axonal path-finding during neurogenesis in *Drosophila*, reflecting its role as a regulator of motile activity in neuronal cells.[165] In the mammalian brain, profilin I is expressed at varying levels in different neuronal subtypes and is found both pre- and postsynaptically,[166] where it appears to participate in activity-dependent remodelling of synapse morphology, which requires actin polymerization.[167] The presynaptic protein aczonin binds profilin [168] and postsynaptically both delphilin and gephyrin are profilin ligands.[169,170] Gephyrin competes with profilin for binding to actin, pointing to a connection between glycine and GABA receptors and the dynamics of actin.[171] Interestingly, neuronal translation of profilin mRNA in *Drosophila* appears to be under negative regulation of the homolog to the Fragile X mental retardation protein (FMRP), indicating a connection between the control of actin polymerization and the Fragile X syndrome.[172]

How the activity of profilin I coordinates with profilin II in brain neurons is unclear. In this case profilin II may function quite differently from profilin I. An alternative role for profilin II in neurons was proposed by Gareus et al[173] who showed that its interaction with the endocytic protein dynamin interferes with the binding of endophilin and amphiphysin to dynamin, suggesting that profilin II primarily operates as a regulator of endocytosis. The significance of the distinction in partner-recognition between the two profilin isoforms remains to be clarified. Notably, profilin I depletion by gene disruption is lethal,[20] while the animals can survive without profilin II.[166]

It was recently demonstrated by imaging that after NMDA-receptor activation and subsequent Ca^{2+}-ion influx, a GFP-tagged profilin II accumulates in dendritic spines.[174] This reduced the motility of the spines and stabilized their morphology. The formation of a rather permanent profilin-containing postsynaptic structure is proposed to reflect the initial stages in the establishment of long term memory.[174]

Conclusions

Clearly, cells use complex protein-machineries to organize their microfilament system in response to extracellular signalling. Ultimately it is a matter of strict spatial and temporal control of actin-dependent force-generation.

Profilin has a central role in this context by 'functionalizing' monomeric actin in a precursor form, profilin:actin-ATP, which, through the poly-(L-proline)-binding surface on profilin, is recognized by different polymer-forming systems and subsequently is used to incorporate new actin molecules into growing filaments. Profilin I appears to be unique for this function, making profilin:actin the principal source of actin for filament formation in most, maybe all eukaryotic cells.

Structure analyses and extensive biochemical and cell biological studies in many laboratories during 30 years have led to far-reaching insights into the molecular mechanisms that govern profilin function. What must come next is efforts to unravel the role of profilin in polyphosphoinositide metabolism, the possible regulation of profilin by phosphorylation, the mechanisms bringing profilin:actin to polymer-forming sites at the cell edge, and the function of profilin and profilin:actin in the nucleus.

Acknowledgements

We gratefully recognize a long standing collaboration with Clarence E. Schutt during which he has generously shared his insights in structural biology. We also thank Herwig Schüler for many valuable discussions and for communicating data on *Plasmodium* profilin before their publication, and Staffan Grenklo for stimulating discussions and collaboration. The Swedish Research Council and the Cancer Foundation are acknowledged for financial support.

References

1. Lindberg MU. Crystallization from calf spleen of two inhibitors of deoxyribonuclease I. J Biol Chem 1966; 241(5):1246-1249.
2. Lindberg U. Studies on the complex formation between deoxyribonuclease I and spleen inhibitor II. Biochemistry 1967; 6(1):343-347.
3. Lazarides E, Lindberg U. Actin is the naturally occurring inhibitor of deoxyribonuclease I. Proc Natl Acad Sci USA 1974; 71(12):4742-4746.
4. Carlsson L, Nystrom LE, Lindberg U et al. Crystallization of a nonmuscle actin. J Mol Biol 1976; 105(3):353-366.
5. Carlsson L, Nystrom LE, Sundkvist I et al. Actin polymerizability is influenced by profilin, a low molecular weight protein in nonmuscle cells. J Mol Biol 1977; 115(3):465-483.
6. Tilney LG. The role of nonmuscle cell motility. In: Inoué S, Stephens RE, eds. Molecules and Cell Movement. New York: Raven Press, 1975:339-388.
7. Tilney LG. The polymerization of actin. III. Aggregates of nonfilamentous actin and its associated proteins: A storage form of actin. J Cell Biol 1976; 69(1):73-89.
8. Tilney LG, Bonder EM, Coluccio LM et al. Actin from Thyone sperm assembles on only one end of an actin filament: A behavior regulated by profilin. J Cell Biol 1983; 97(1):112-124.
9. Reichstein E, Korn ED. Acanthamoeba profilin. A protein of low molecular weight from Acanpthamoeba castellanii that inhibits actin nucleation. J Biol Chem 1979; 254(13):6174-6179.
10. Tobacman LS, Korn ED. The regulation of actin polymerization and the inhibition of monomeric actin ATPase activity by Acanthamoeba profilin. J Biol Chem 1982; 257(8):4166-4170.
11. Markey F, Persson T, Lindberg U. Characterization of platelet extracts before and after stimulation with respect to the possible role of profilactin as microfilament precursor. Cell 1981; 23(1):145-153.
12. Pollard TD, Mooseker MS. Direct measurement of actin polymerization rate constants by electron microscopy of actin filaments nucleated by isolated microvillus cores. J Cell Biol 1981; 88(3):654-659.
13. Markey F, Larsson H, Weber K et al. Nucleation of actin polymerization from profilactin. Opposite effects of different nuclei. Biochim Biophys Acta 1982; 704(1):43-51.
14. Pollard TD, Cooper JA. Quantitative analysis of the effect of Acanthamoeba profilin on actin filament nucleation and elongation. Biochemistry 1984; 23(26):6631-6641.
15. Pring M, Weber A, Bubb MR. Profilin-actin complexes directly elongate actin filaments at the barbed end. Biochemistry 1992; 31(6):1827-1836.
16. Schutt CE, Myslik JC, Rozycki MD et al. The structure of crystalline profilin-beta-actin. Nature 1993; 365(6449):810-816.
17. Cedergren-Zeppezauer ES, Goonesekere NC, Rozycki MD et al. Crystallization and structure determination of bovine profilin at 2.0 A resolution. J Mol Biol 1994; 240(5):459-475.
18. Balasubramanian MK, Hirani BR, Burke JD et al. The Schizosaccharomyces pombe cdc3+ gene encodes a profilin essential for cytokinesis. J Cell Biol 1994; 125(6):1289-1301.
19. Cooley L, Verheyen E, Ayers K. chickadee encodes a profilin required for intercellular cytoplasm transport during Drosophila oogenesis. Cell 1992; 69(1):173-184.
20. Witke W, Sutherland JD, Sharpe A et al. Profilin I is essential for cell survival and cell division in early mouse development. Proc Natl Acad Sci USA 2001; 98(7):3832-3836.

21. Kandasamy MK, McKinney EC, Meagher RB. Plant profilin isovariants are distinctly regulated in vegetative and reproductive tissues. Cell Motil Cytoskeleton 2002; 52(1):22-32.
22. Witke W. The role of profilin complexes in cell motility and other cellular processes. Trends Cell Biol 2004; 14(8):461-469.
23. Honore B, Madsen P, Andersen AH et al. Cloning and expression of a novel human profilin variant, profilin II. FEBS Lett 1993; 330(2):151-155.
24. Witke W, Podtelejnikov AV, Di Nardo A et al. In mouse brain profilin I and profilin II associate with regulators of the endocytic pathway and actin assembly. EMBO J 1998; 17(4):967-976.
25. Di Nardo A, Gareus R, Kwiatkowski D et al. Alternative splicing of the mouse profilin II gene generates functionally different profilin isoforms. J Cell Sci 2000; 113(Pt 21):3795-3803.
26. Lambrechts A, Braun A, Jonckheere V et al. Profilin II is alternatively spliced, resulting in profilin isoforms that are differentially expressed and have distinct biochemical properties. Mol Cell Biol 2000; 20(21):8209-8219.
27. Machesky LM, Cole NB, Moss B et al. Vaccinia virus expresses a novel profilin with a higher affinity for polyphosphoinositides than actin. Biochemistry 1994; 33(35):10815-10824.
28. Braun A, Aszodi A, Hellebrand H et al. Genomic organization of profilin-III and evidence for a transcript expressed exclusively in testis. Gene 2002; 283(1-2):219-225.
29. Obermann H, Raabe I, Balvers M et al. Novel testis-expressed profilin IV associated with acrosome biogenesis and spermatid elongation. Mol Hum Reprod 2005; 11(1):53-64.
30. Radauer C, Breiteneder H. Pollen allergens are restricted to few protein families and show distinct patterns of species distribution. J Allergy Clin Immunol 2006; 117(1):141-147.
31. Valenta R, Duchene M, Ebner C et al. Profilins constitute a novel family of functional plant pan-allergens. J Exp Med 1992; 175(2):377-385.
32. Yarovinsky F, Zhang D, Andersen JF et al. TLR11 activation of dendritic cells by a protozoan profilin-like protein. Science 2005; 308(5728):1626-1629.
33. Rothkegel M, Mayboroda O, Rohde M et al. Plant and animal profilins are functionally equivalent and stabilize microfilaments in living animal cells. J Cell Sci 1996; 109(Pt 1):83-90.
34. Karakesisoglou I, Schleicher M, Gibbon BC et al. Plant profilins rescue the aberrant phenotype of profilin-deficient Dictyostelium cells. Cell Motil Cytoskeleton 1996; 34(1):36-47.
35. Eads JC, Mahoney NM, Vorobiev S et al. Structure determination and characterization of Saccharomyces cerevisiae profilin. Biochemistry 1998; 37(32):11171-11181.
36. Fedorov AA, Magnus KA, Graupe MH et al. X-ray structures of isoforms of the actin-binding protein profilin that differ in their affinity for phosphatidylinositol phosphates. Proc Natl Acad Sci USA 1994; 91(18):8636-8640.
37. Nodelman IM, Bowman GD, Lindberg U et al. X-ray structure determination of human profilin II: A comparative structural analysis of human profilins. J Mol Biol 1999; 294(5):1271-1285.
38. Thorn KS, Christensen HE, Shigeta R et al. The crystal structure of a major allergen from plants. Structure 1997; 5(1):19-32.
39. Hajkova L, Bjorkegren Sjogren C, Korenbaum E et al. Characterization of a mutant profilin with reduced actin-binding capacity: Effects in vitro and in vivo. Exp Cell Res 1997; 234(1):66-77.
40. Korenbaum E, Nordberg P, Bjorkegren-Sjogren C et al. The role of profilin in actin polymerization and nucleotide exchange. Biochemistry 1998; 37(26):9274-9283.
41. Qian C, Zhang Q, Wang X et al. Structure of the adaptor protein p14 reveals a profilin-like fold with distinct function. J Mol Biol 2005; 347(2):309-321.
42. Rossi V, Banfield DK, Vacca M et al. Longins and their longin domains: Regulated SNAREs and multifunctional SNARE regulators. Trends Biochem Sci 2004; 29(12):682-688.
43. Lindberg U, Schutt CE, Hellsten E et al. The use of poly(L-proline)-Sepharose in the isolation of profilin and profilactin complexes. Biochim Biophys Acta 1988; 967(3):391-400.
44. Schüler H, Karlsson R, Lindberg U. Purification of nonmuscle actin. In: Celis J, ed. Cell Biology: A Laboratory Handbook, Vol I. 3th ed. London: Elsevier Science, 2006:165-171.
45. Reinhard M, Giehl K, Abel K et al. The proline-rich focal adhesion and microfilament protein VASP is a ligand for profilins. EMBO J 1995; 14(8):1583-1589.
46. Mahoney NM, Janmey PA, Almo SC. Structure of the profilin-poly-L-proline complex involved in morphogenesis and cytoskeletal regulation. Nat Struct Biol 1997; 4(11):953-960.
47. Mahoney NM, Rozwarski DA, Fedorov E et al. Profilin binds proline-rich ligands in two distinct amide backbone orientations. Nat Struct Biol 1999; 6(7):666-671.
48. Bjorkegren C, Rozycki M, Schutt CE et al. Mutagenesis of human profilin locates its poly(L-proline)-binding site to a hydrophobic patch of aromatic amino acids. FEBS Lett 1993; 333(1-2):123-126.

49. Lambrechts A, Jonckheere V, Dewitte D et al. Mutational analysis of human profilin I reveals a second PI(4,5)-P2 binding site neighbouring the poly(L-proline) binding site. BMC Biochem 2002; 3:12.
50. Skare P, Karlsson R. Evidence for two interaction regions for phosphatidylinositol(4,5)-bisphosphate on mammalian profilin I. FEBS Lett 2002; 522(1-3):119-124.
51. Malm B, Larsson H, Lindberg U. The profilin-actin complex: Further characterization of profilin and studies on the stability of the complex. J Muscle Res Cell Motil 1983; 4(5):569-588.
52. Kasina S, Rizwani W, Radhika KV et al. Nitration of profilin effects its interaction with poly (L-proline) and actin. J Biochem (Tokyo) 2005; 138(6):687-695.
53. Bjorkegren-Sjogren C, Korenbaum E, Nordberg P et al. Isolation and characterization of two mutants of human profilin I that do not bind poly(L-proline). FEBS Lett 1997; 418(3):258-264.
54. Bjorkegren C. Profilin in cell motility and signal transduction /Ph.D.-thesis, ISBN 91-7153-646-9. Stockholm: Department of Cell Biology, Stockholm University, 1997.
55. Raghunathan V, Mowery P, Rozycki M et al. Structural changes in profilin accompany its binding to phosphatidylinositol, 4,5-bisphosphate. FEBS Lett 1992; 297(1-2):46-50.
56. Lassing I, Lindberg U. Specific interaction between phosphatidylinositol 4,5-bisphosphate and profilactin. Nature 1985; 314(6010):472-474.
57. Kaiser DA, Pollard TD. Characterization of actin and poly-L-proline binding sites of Acanthamoeba profilin with monoclonal antibodies and by mutagenesis. J Mol Biol 1996; 256(1):89-107.
58. De Corte V, Gettemans J, Vandekerckhove J. Phosphatidylinositol 4,5-bisphosphate specifically stimulates PP60(c-src) catalyzed phosphorylation of gelsolin and related actin-binding proteins. FEBS Lett 1997; 401(2-3):191-196.
59. Hansson A, Skoglund G, Lassing I et al. Protein kinase C-dependent phosphorylation of profilin is specifically stimulated by phosphatidylinositol bisphosphate (PIP2). Biochem Biophys Res Commun 1988; 150(2):526-531.
60. Sathish K, Padma B, Munugalavadla V et al. Phosphorylation of profilin regulates its interaction with actin and poly (L-proline). Cell Signal 2004; 16(5):589-596.
61. Singh SS, Chauhan A, Murakami N et al. Phosphoinositide-dependent in vitro phosphorylation of profilin by protein kinase C. Phospholipid specificity and localization of the phosphorylation site. Recept Signal Transduct 1996; 6(2):77-86.
62. Limmongkon A, Giuliani C, Valenta R et al. MAP kinase phosphorylation of plant profilin. Biochem Biophys Res Commun 2004; 324(1):382-386.
63. Guillen G, Valdes-Lopez V, Noguez R et al. Profilin in Phaseolus vulgaris is encoded by two genes (only one expressed in root nodules) but multiple isoforms are generated in vivo by phosphorylation on tyrosine residues. Plant J 1999; 19(5):497-508.
64. Kabsch W, Mannherz HG, Suck D et al. Atomic structure of the actin: DNase I complex. Nature 1990; 347(6288):37-44.
65. McLaughlin PJ, Gooch JT, Mannherz HG et al. Structure of gelsolin segment 1-actin complex and the mechanism of filament severing. Nature 1993; 364(6439):685-692.
66. Vorobiev S, Strokopytov B, Drubin DG et al. The structure of nonvertebrate actin: Implications for the ATP hydrolytic mechanism. Proc Natl Acad Sci USA 2003; 100(10):5760-5765.
67. Chik JK, Lindberg U, Schutt CE. The structure of an open state of beta-actin at 2.65 A resolution. J Mol Biol 1996; 263(4):607-623.
68. Otterbein LR, Graceffa P, Dominguez R. The crystal structure of uncomplexed actin in the ADP state. Science 2001; 293(5530):708-711.
69. Mockrin SC, Korn ED. Acanthamoeba profilin interacts with G-actin to increase the rate of exchange of actin-bound adenosine 5'-triphosphate. Biochemistry 1980; 19(23):5359-5362.
70. Nyman T, Page R, Schutt CE et al. A cross-linked profilin-actin heterodimer interferes with elongation at the fast-growing end of F-actin. J Biol Chem 2002; 277(18):15828-15833.
71. Schüler H, Karlsson R, Schutt CE et al. The connection between actin ATPase and polymerization. In: Bittar EE, ed. Advances in Molcecular and Cellular Biology, Vol 37. Amsterdam: Elsevier, 2006.
72. Hajkova L, Nyman T, Lindberg U et al. Effects of cross-linked profilin:beta/gamma-actin on the dynamics of the microfilament system in cultured cells. Exp Cell Res 2000; 256(1):112-121.
73. Suetsugu S, Miki H, Takenawa T. The essential role of profilin in the assembly of actin for microspike formation. EMBO J 1998; 17(22):6516-6526.
74. Suetsugu S, Miki H, Takenawa T. Distinct roles of profilin in cell morphological changes: Microspikes, membrane ruffles, stress fibers, and cytokinesis. FEBS Lett 1999; 457(3):470-474.
75. Janke J, Schluter K, Jandrig B et al. Suppression of tumorigenicity in breast cancer cells by the microfilament protein profilin 1. J Exp Med 2000; 191(10):1675-1686.

76. Roy P, Jacobson K. Overexpression of profilin reduces the migration of invasive breast cancer cells. Cell Motil Cytoskeleton 2004; 57(2):84-95.
77. Wittenmayer N, Jandrig B, Rothkegel M et al. Tumor suppressor activity of profilin requires a functional actin binding site. Mol Biol Cell 2004; 15(4):1600-1608.
78. Chaudhary A, Chen J, Gu QM et al. Probing the phosphoinositide 4,5-bisphosphate binding site of human profilin I. Chem Biol 1998; 5(5):273-281.
79. Haarer BK, Petzold AS, Brown SS. Mutational analysis of yeast profilin. Mol Cell Biol 1993; 13(12):7864-7873.
80. Kovar DR, Drobak BK, Collings DA et al. The characterization of ligand-specific maize (Zea mays) profilin mutants. Biochem J 2001; 358(Pt 1):49-57.
81. Lu PJ, Shieh WR, Rhee SG et al. Lipid products of phosphoinositide 3-kinase bind human profilin with high affinity. Biochemistry 1996; 35(44):14027-14034.
82. Sohn RH, Chen J, Koblan KS et al. Localization of a binding site for phosphatidylinositol 4,5-bisphosphate on human profilin. J Biol Chem 1995; 270(36):21114-21120.
83. Hartwig JH, Bokoch GM, Carpenter CL et al. Thrombin receptor ligation and activated Rac uncap actin filament barbed ends through phosphoinositide synthesis in permeabilized human platelets. Cell 1995; 82(4):643-653.
84. Lassing I, Lindberg U. Polyphosphoinositide synthesis in platelets stimulated with low concentrations of thrombin is enhanced before the activation of phospholipase C. FEBS Lett 1990; 262(2):231-233.
85. Carlsson L, Markey F, Blikstad I et al. Reorganization of actin in platelets stimulated by thrombin as measured by the DNase I inhibition assay. Proc Natl Acad Sci USA 1979; 76(12):6376-6380.
86. Lassing I, Lindberg U. Evidence that the phosphatidylinositol cycle is linked to cell motility. Exp Cell Res 1988; 174(1):1-15.
87. Goldschmidt-Clermont PJ, Kim JW, Machesky LM et al. Regulation of phospholipase C-gamma 1 by profilin and tyrosine phosphorylation. Science 1991; 251(4998):1231-1233.
88. Goldschmidt-Clermont PJ, Machesky LM, Baldassare JJ et al. The actin-binding protein profilin binds to PIP2 and inhibits its hydrolysis by phospholipase C. Science 1990; 247(4950):1575-1578.
89. Singh SS, Chauhan A, Murakami N et al. Profilin and gelsolin stimulate phosphatidylinositol 3-kinase activity. Biochemistry 1996; 35(51):16544-16549.
90. Han J, Luby-Phelps K, Das B et al. Role of substrates and products of PI 3-kinase in regulating activation of Rac-related guanosine triphosphatases by Vav. Science 1998; 279(5350):558-560.
91. Jaffe AB, Hall A. Rho GTPases: Biochemistry and biology. Annu Rev Cell Dev Biol 2005; 21:247-269.
92. Ren XD, Schwartz MA. Regulation of inositol lipid kinases by Rho and Rac. Curr Opin Genet Dev 1998; 8(1):63-67.
93. Ostrander DB, Gorman JA, Carman GM. Regulation of profilin localization in Saccharomyces cerevisiae by phosphoinositide metabolism. J Biol Chem 1995; 270(45):27045-27050.
94. Buss F, Temm-Grove C, Henning S et al. Distribution of profilin in fibroblasts correlates with the presence of highly dynamic actin filaments. Cell Motil Cytoskeleton 1992; 22(1):51-61.
95. Mayboroda O, Schluter K, Jockusch BM. Differential colocalization of profilin with microfilaments in PtK2 cells. Cell Motil Cytoskeleton 1997; 37(2):166-177.
96. Giesemann T, Rathke-Hartlieb S, Rothkegel M et al. A role for polyproline motifs in the spinal muscular atrophy protein SMN. Profilins bind to and colocalize with smn in nuclear gems. J Biol Chem 1999; 274(53):37908-37914.
97. Skare P, Kreivi JP, Bergstrom A, Karlsson R. Profilin I colocalizes with speckles and Cajal bodies: A possible role in pre-mRNA splicing. Exp Cell Res 2003; 286(1):12-21.
98. Pollard TD, Borisy GG. Cellular motility driven by assembly and disassembly of actin filaments. Cell 2003; 112(4):453-465.
99. Machesky LM, Atkinson SJ, Ampe C et al. Purification of a cortical complex containing two unconventional actins from Acanthamoeba by affinity chromatography on profilin-agarose. J Cell Biol 1994; 127(1):107-115.
100. Mullins RD, Kelleher JF, Xu J et al. Arp2/3 complex from Acanthamoeba binds profilin and cross-links actin filaments. Mol Biol Cell 1998; 9(4):841-852.
101. Machesky LM, Mullins RD, Higgs HN et al. Scar, a WASp-related protein, activates nucleation of actin filaments by the Arp2/3 complex. Proc Natl Acad Sci USA 1999; 96(7):3739-3744.
102. Rohatgi R, Ma L, Miki H et al. The interaction between N-WASP and the Arp2/3 complex links Cdc42-dependent signals to actin assembly. Cell 1999; 97(2):221-231.
103. Winter D, Lechler T, Li R. Activation of the yeast Arp2/3 complex by Bee1p, a WASP-family protein. Curr Biol 1999; 9(9):501-504.

104. Uruno T, Liu J, Zhang P et al. Activation of Arp2/3 complex-mediated actin polymerization by cortactin. Nat Cell Biol 2001; 3(3):259-266.
105. Frischknecht F, Way M. Surfing pathogens and the lessons learned for actin polymerization. Trends Cell Biol 2001; 11(1):30-38.
106. Welch MD, Rosenblatt J, Skoble J et al. Interaction of human Arp2/3 complex and the Listeria monocytogenes ActA protein in actin filament nucleation. Science 1998; 281(5373):105-108.
107. Geese M, Schluter K, Rothkegel M et al. Accumulation of profilin II at the surface of Listeria is concomitant with the onset of motility and correlates with bacterial speed. J Cell Sci 2000; 113(Pt 8):1415-1426.
108. Grenklo S, Geese M, Lindberg U et al. A crucial role for profilin-actin in the intracellular motility of Listeria monocytogenes. EMBO Rep 2003; 4(5):523-529.
109. Loisel TP, Boujemaa R, Pantaloni D et al. Reconstitution of actin-based motility of Listeria and Shigella using pure proteins. Nature 1999; 401(6753):613-616.
110. Mullins RD, Heuser JA, Pollard TD. The interaction of Arp2/3 complex with actin: Nucleation, high affinity pointed end capping, and formation of branching networks of filaments. Proc Natl Acad Sci USA 1998; 95(11):6181-6186.
111. Bailly M, Ichetovkin I, Grant W et al. The F-actin side binding activity of the Arp2/3 complex is essential for actin nucleation and lamellipod extension. Curr Biol 2001; 11(8):620-625.
112. Blanchoin L, Amann KJ, Higgs HN et al. Direct observation of dendritic actin filament networks nucleated by Arp2/3 complex and WASP/Scar proteins. Nature 2000; 404(6781):1007-1011.
113. Egile C, Rouiller I, Xu XP et al. Mechanism of filament nucleation and branch stability revealed by the structure of the Arp2/3 complex at actin branch junctions. PLoS Biol 2005; 3(11):e383.
114. Pantaloni D, Boujemaa R, Didry D et al. The Arp2/3 complex branches filament barbed ends: Functional antagonism with capping proteins. Nat Cell Biol 2000; 2(7):385-391.
115. Volkmann N, Amann KJ, Stoilova-McPhie S et al. Structure of Arp2/3 complex in its activated state and in actin filament branch junctions. Science 2001; 293(5539):2456-2459.
116. Robinson RC, Turbedsky K, Kaiser DA et al. Crystal structure of Arp2/3 complex. Science 2001; 294(5547):1679-1684.
117. Svitkina TM, Borisy GG. Arp2/3 complex and actin depolymerizing factor/cofilin in dendritic organization and treadmilling of actin filament array in lamellipodia. J Cell Biol 1999; 145(5):1009-1026.
118. Svitkina TM, Verkhovsky AB, McQuade KM et al. Analysis of the actin-myosin II system in fish epidermal keratocytes: Mechanism of cell body translocation. J Cell Biol 1997; 139(2):397-415.
119. Pollard TD, Blanchoin L, Mullins RD. Molecular mechanisms controlling actin filament dynamics in nonmuscle cells. Annu Rev Biophys Biomol Struct 2000; 29:545-576.
120. Hoglund AS, Karlsson R, Arro E et al. Visualization of the peripheral weave of microfilaments in glia cells. J Muscle Res Cell Motil 1980; 1(2):127-146.
121. Small JV, Rinnerthaler G, Hinssen H. Organization of actin meshworks in cultured cells: The leading edge. Cold Spring Harb Symp Quant Biol 1982; 46(Pt 2):599-611.
122. Lindberg U, Hoglund AS, Karlsson R. On the ultrastructural organization of the microfilament system and the possible role of profilactin. Biochimie 1981; 63(4):307-323.
123. Svitkina TM, Verkhovsky AB, Borisy GG. Improved procedures for electron microscopic visualization of the cytoskeleton of cultured cells. J Struct Biol 1995; 115(3):290-303.
124. Le Clainche C, Pantaloni D, Carlier MF. ATP hydrolysis on actin-related protein 2/3 complex causes debranching of dendritic actin arrays. Proc Natl Acad Sci USA 2003; 100(11):6337-6342.
125. Ichetovkin I, Grant W, Condeelis J. Cofilin produces newly polymerized actin filaments that are preferred for dendritic nucleation by the Arp2/3 complex. Curr Biol 2002; 12(1):79-84.
126. Blanchoin L, Pollard TD, Hitchcock-DeGregori SE. Inhibition of the Arp2/3 complex-nucleated actin polymerization and branch formation by tropomyosin. Curr Biol 2001; 11(16):1300-1304.
127. DesMarais V, Ghosh M, Eddy R et al. Cofilin takes the lead. J Cell Sci 2005; 118(Pt 1):19-26.
128. DesMarais V, Ichetovkin I, Condeelis J et al. Spatial regulation of actin dynamics: A tropomyosin-free, actin-rich compartment at the leading edge. J Cell Sci 2002; 115(Pt 23):4649-4660.
129. Hillberg L, Zhao-Rathje LS, Nyåkern-Meazza M et al. Tropomyosins are present in lamellipodia of motile cells. Eur J of Cell Biol 2006, (in press).
130. Di Nardo A, Cicchetti G, Falet H et al. Arp2/3 complex-deficient mouse fibroblasts are viable and have normal leading-edge actin structure and function. Proc Natl Acad Sci USA 2005; 102(45):16263-16268.
131. Yarar D, D'Alessio JA, Jeng RL et al. Motility determinants in WASP family proteins. Mol Biol Cell 2002; 13(11):4045-4059.

132. Rottner K, Behrendt B, Small JV et al. VASP dynamics during lamellipodia protrusion. Nat Cell Biol 1999; 1(5):321-322.
133. Zigmond SH. Formin-induced nucleation of actin filaments. Curr Opin Cell Biol 2004; 16(1):99-105.
134. Higashida C, Miyoshi T, Fujita A et al. Actin polymerization-driven molecular movement of mDial in living cells. Science 2004; 303(5666):2007-2010.
135. Kovar DR. Molecular details of formin-mediated actin assembly. Curr Opin Cell Biol 2006; 18(1):11-17.
136. Romero S, Le Clainche C, Didry D et al. Formin is a processive motor that requires profilin to accelerate actin assembly and associated ATP hydrolysis. Cell 2004; 119(3):419-429.
137. Geese M, Loureiro JJ, Bear JE et al. Contribution of Ena/VASP proteins to intracellular motility of listeria requires phosphorylation and proline-rich core but not F-actin binding or multimerization. Mol Biol Cell 2002; 13(7):2383-2396.
138. Grenklo S, Johansson T, Bertilson L et al. Anti-actin antibodies generated against profilin:actin distinguish between nonfilamentous and filamentous actin, and label cultured cells in a dotted pattern. Eur J Cell Biol 2004; 83(8):413-423.
139. Dickinson RB, Purich DL. Clamped-filament elongation model for actin-based motors. Biophys J 2002; 82(2):605-617.
140. Dickinson RB, Southwick FS, Purich DL. A direct-transfer polymerization model explains how the multiple profilin-binding sites in the actoclampin motor promote rapid actin-based motility. Arch Biochem Biophys 2002; 406(2):296-301.
141. Ho HY, Rohatgi R, Lebensohn AM et al. Toca-1 mediates Cdc42-dependent actin nucleation by activating the N-WASP-WIP complex. Cell 2004; 118(2):203-216.
142. Steffen A, Rottner K, Ehinger J et al. Sra-1 and Nap1 link Rac to actin assembly driving lamellipodia formation. EMBO J 2004; 23(4):749-759.
143. Disanza A, Steffen A, Hertzog M et al. Actin polymerization machinery: The finish line of signaling networks, the starting point of cellular movement. Cell Mol Life Sci 2005; 62(9):955-970.
144. Stradal TE, Rottner K, Disanza A et al. Regulation of actin dynamics by WASP and WAVE family proteins. Trends Cell Biol 2004; 14(6):303-311.
145. Allen PG. Actin filament uncapping localizes to ruffling lamellae and rocketing vesicles. Nat Cell Biol 2003; 5(11):972-979.
146. Disanza A, Carlier MF, Stradal TE et al. Eps8 controls actin-based motility by capping the barbed ends of actin filaments. Nat Cell Biol 2004; 6(12):1180-1188.
147. Falck S, Paavilainen VO, Wear MA et al. Biological role and structural mechanism of twinfilin-capping protein interaction. EMBO J 2004; 23(15):3010-3019.
148. Sun HQ, Yamamoto M, Mejillano M et al. Gelsolin, a multifunctional actin regulatory protein. J Biol Chem 1999; 274(47):33179-33182.
149. Wear MA, Cooper JA. Capping protein: New insights into mechanism and regulation. Trends Biochem Sci 2004; 29(8):418-428.
150. Yang C, Pring M, Wear MA et al. Mammalian CARMIL inhibits actin filament capping by capping protein. Dev Cell 2005; 9(2):209-221.
151. Nyakern-Meazza M, Narayan K, Schutt CE et al. Tropomyosin and gelsolin cooperate in controlling the microfilament system. J Biol Chem 2002; 277(32):28774-28779.
152. Evangelista M, Pruyne D, Amberg DC et al. Formins direct Arp2/3-independent actin filament assembly to polarize cell growth in yeast. Nat Cell Biol 2002; 4(3):260-269.
153. Paavilainen VO, Bertling E, Falck S et al. Regulation of cytoskeletal dynamics by actin-monomer-binding proteins. Trends Cell Biol 2004; 14(7):386-394.
154. Zicha D, Dobbie IM, Holt MR et al. Rapid actin transport during cell protrusion. Science 2003; 300(5616):142-145.
155. Berg JS, Cheney RE. Myosin-X is an unconventional myosin that undergoes intrafilopodial motility. Nat Cell Biol 2002; 4(3):246-250.
156. Zhang H, Berg JS, Li Z et al. Myosin-X provides a motor-based link between integrins and the cytoskeleton. Nat Cell Biol 2004; 6(6):523-531.
157. Briese M, Esmaeili B, Sattelle DB. Is spinal muscular atrophy the result of defects in motor neuron processes? Bioessays 2005; 27(9):946-957.
158. Monani UR. Spinal muscular atrophy: A deficiency in a ubiquitous protein; a motor neuron-specific disease. Neuron 2005; 48(6):885-896.
159. Boronenkov IV, Loijens JC, Umeda M et al. Phosphoinositide signaling pathways in nuclei are associated with nuclear speckles containing pre-mRNA processing factors. Mol Biol Cell 1998; 9(12):3547-3560.

160. Osborne SL, Thomas CL, Gschmeissner S et al. Nuclear PtdIns(4,5)P2 assembles in a mitotically regulated particle involved in pre-mRNA splicing. J Cell Sci 2001; 114(Pt 13):2501-2511.
161. Bettinger BT, Gilbert DM, Amberg DC. Actin up in the nucleus. Nat Rev Mol Cell Biol 2004; 5(5):410-415.
162. Pederson T, Aebi U. Nuclear actin extends, with no contraction in sight. Mol Biol Cell 2005; 16(11):5055-5060.
163. Stuven T, Hartmann E, Gorlich D. Exportin 6: A novel nuclear export receptor that is specific for profilin.actin complexes. EMBO J 2003; 22(21):5928-5940.
164. Lederer M, Jockusch BM, Rothkegel M. Profilin regulates the activity of p42POP, a novel Myb-related transcription factor. J Cell Sci 2005; 118(Pt 2):331-341.
165. Wills Z, Marr L, Zinn K et al. Profilin and the Abl tyrosine kinase are required for motor axon outgrowth in the Drosophila embryo. Neuron 1999; 22(2):291-299.
166. Neuhoff H, Sassoe-Pognetto M, Panzanelli P et al. The actin-binding protein profilin I is localized at synaptic sites in an activity-regulated manner. Eur J Neurosci 2005; 21(1):15-25.
167. Fischer M, Kaech S, Knutti D et al. Rapid actin-based plasticity in dendritic spines. Neuron 1998; 20(5):847-854.
168. Wang X, Kibschull M, Laue MM et al. Aczonin, a 550-kD putative scaffolding protein of presynaptic active zones, shares homology regions with Rim and Bassoon and binds profilin. J Cell Biol 1999; 147(1):151-162.
169. Mammoto A, Sasaki T, Asakura T et al. Interactions of drebrin and gephyrin with profilin. Biochem Biophys Res Commun 1998; 243(1):86-89.
170. Miyagi Y, Yamashita T, Fukaya M et al. Delphilin: A novel PDZ and formin homology domain-containing protein that synaptically colocalizes and interacts with glutamate receptor delta 2 subunit. J Neurosci 2002; 22(3):803-814.
171. Giesemann T, Schwarz G, Nawrotzki R et al. Complex formation between the postsynaptic scaffolding protein gephyrin, profilin, and Mena: A possible link to the microfilament system. J Neurosci 2003; 23(23):8330-8339.
172. Reeve SP, Bassetto L, Genova GK et al. The Drosophila fragile X mental retardation protein controls actin dynamics by directly regulating profilin in the brain. Curr Biol 2005; 15(12):1156-1163.
173. Gareus R, Di Nardo A, Rybin V et al. Mouse profilin 2 regulates endocytosis and competes with SH3-ligand binding to dynamin 1. J Biol Chem 2006; 281(5):2803-2811.
174. Ackermann M, Matus A. Activity-induced targeting of profilin and stabilization of dendritic spine morphology. Nat Neurosci 2003; 6(11):1194-1200.

CHAPTER

Srv2/Cyclase-Associated Protein (CAP): A Multi-Functional Recycling Center for Actin Monomers and Cofilin

Bruce L. Goode*

Abstract

Srv2/cyclase-associated protein (CAP) is a ubiquitously expressed actin monomer binding protein required for proper organization and rapid remodeling of cellular actin networks. CAP catalyzes the dissociation of cofilin-bound ADP-actin complexes, elevating cofilin levels available for filament disassembly. In addition, CAP and profilin promote exchange of nucleotide (ATP for ADP) on G-actin, then CAP releases profilin-bound ATP-G-actin to replenish the actin monomer pool. These functions are highly conserved, as expression of animal and plant CAPs complement cellular defects of yeast *cap* mutants. Unlike most actin monomer binding proteins, CAP oligomerizes, likely into hexamers. Within the high molecular weight complex formed, the C-terminal half of each CAP molecule binds one actin monomer. In addition, the N-terminus of CAP binds to cofilin-G-actin complexes and the middle region binds to Abp1 and profilin. Abp1 tethers CAP to filamentous actin networks. Cofilin and profilin function together with CAP to accelerate actin turnover, through a series of actin monomer handoffs guided by the changing nucleotide state of actin. Thus, the emerging view of CAP function is that it serves as a large molecular hub where multiple actin binding proteins interact to recycle actin monomers and cofilin. This macromolecular complex plays a key role in remodeling the actin cytoskeleton during events such as endocytosis, cell polarity, and cell motility.

Introduction

The actin cytoskeleton plays a crucial role in directing alterations in cell shape, polarity, and infrastructure in response to cues. Proper remodeling of the actin cytoskeleton is required for a wide range of cellular processes, including endocytosis, intracellular transport, cytokinesis, and cell motility. These dynamic events depend on mechanisms both for nucleating new actin filament assembly and for rapidly dismantling existing filaments and recycling their subunits to replenish the actin monomer pool. The rate-limiting step in filament disassembly is the dissociation of subunits from pointed ends of filaments. This step is accelerated up to 20-100 fold by the addition of cofilin/ADF, which severs filaments and/or promotes dissociation of subunits from pointed ends of filaments.[1] In animal cells, cofilin/ADF activity is tightly controlled by PIP_2 interactions and by phosphorylation.[2] Transient changes in cofilin/ADF activity can drastically change actin cytoskeleton organization and dynamics. Thus, modulation of cofilin/ADF activity is a key control point in actin remodeling.

*Bruce L. Goode—Rosenstiel Basic Medical Sciences Research Center, Brandeis University, 415 South Street, Waltham, Massachusetts 02454 U.S.A. Email: goode@brandeis.edu

Actin-Monomer-Binding Proteins, edited by Pekka Lappalainen. ©2007 Landes Bioscience and Springer Science+Business Media.

Despite its central role in driving actin filament turnover, cofilin/ADF does not perform this function alone in cells. At least four other ubiquitously expressed actin-binding proteins contribute strongly to cofilin/ADF-dependent actin turnover: Aip1, profilin, twinfilin, and cyclase-associated protein (CAP). Aip1 binds to F-actin and cofilin/ADF and induces cofilin/ADF-dependent net depolymerization of filaments.[3] However, Aip1 function has not been linked to CAP, and thus is not discussed further here. Profilin binds directly to CAP, and its functional relationship with CAP is addressed below. Other chapters in this volume are dedicated to the biochemical mechanisms of profilin and twinfilin and discuss their roles in actin turnover.

CAP, also called Srv2, was first identified in *S. cerevisiae* as a suppressor of a hyper-activated *ras2* (Val19) allele and as a 70kDa protein associated with adenylyl cyclase.[4-6] These and other early studies showed that yeast *cap* mutants are defective in both Ras signaling to adenylyl cyclase and in actin organization and cell morphology.[7,8] Further, mutational analyses showed that these two phenotypes are genetically separable. Since then, CAP homologues have been identified in a wide range of animal, fungal, and plant species, and in all cases CAP localizes to the actin cytoskeleton.[9] Although the link to adenylyl cyclase does not appear to be conserved in CAP outside of fungal species, CAP-defective cells in *Dictyostelium*, *Drosophila*, plants, and mammals exhibit severe defects in actin organization, as well as endocytosis, cell morphogenesis, and cell motility.[10-15] Further, CAP is expressed in all tissue and cell types examined, and animal and plant CAP homologues complement the cellular growth and actin defects of *cap* mutant yeast.[10,14,16,17] Thus, CAP has a highly conserved function in the actin cytoskeleton.

Until recently the mechanism by which CAP regulates actin dynamics was unknown. Early studies showed that the C-terminus of CAP binds to G-actin and has a central role in regulating actin organization in vivo.[7,10,16-19] This led investigators to propose that CAP sequesters actin monomers and thereby promotes F-actin depolymerization. However, recent studies on CAP activity in the presence of its in vivo cofactors cofilin and profilin have provided new insights into the CAP mechanism (Fig. 1). These studies show that CAP promotes actin turnover by (a) recycling cofilin from ADP-actin monomers and (b) promoting nucleotide exchange on actin monomers, thus replenishing the pool of cofilin available for filament depolymerization and actin monomers available for actin assembly.[15,20-22] The focus of this review is on these and other recent advances in understanding CAP structure, interactions with other actin binding proteins, and the mechanism of its cellular function. For an excellent review of the earlier CAP literature readers are referred elsewhere.[9]

CAP Atomic Structure and Assembly into Complexes

CAP fractionates as part of a high molecular weight complex in cell lysates from animal and fungal species.[18,23,24] The relative enormity of these complexes (~600kDa) led to speculation that they result from a combination of dimerization of CAP (57kDa) and its association with adenylyl cyclase (200kDa). However, subsequent studies showed that purified native yeast CAP is comprised entirely of CAP and actin, in a 1:1 molar ratio, with no other proteins detected[21] (Fig. 1B). Further, the hydrodynamic properties (sedimentation coefficient and Stokes radius) of this purified complex indicated a MW of 600kDa, matching CAP complex properties in crude cell lysates. Thus, native CAP complexes appear to be heterododecamers, comprised of 6 molecules of CAP (57kDa) and 6 molecules of actin (43kDa). Consistent with these observations, gel filtration analysis of full-length recombinant CAP (in the absence of actin) suggests that CAP oligomerizes into hexamers or a higher order species.[20,25] These data also demonstrate that CAP oligomerization does not require actin and is mediated by CAP-CAP interactions.

Electron micrographs of purified CAP-actin complexes reveal dimensions consistent with its Stokes radius.[21] These images provide the first low-resolution glimpses of intact complex structure (Fig. 1A). Although molecular assignments for CAP and actin within these structures have not been made, there is a central core region consisting of multiple smaller lobes surrounded by up to six larger globules. In the model for CAP function (Fig. 1C), the outer larger

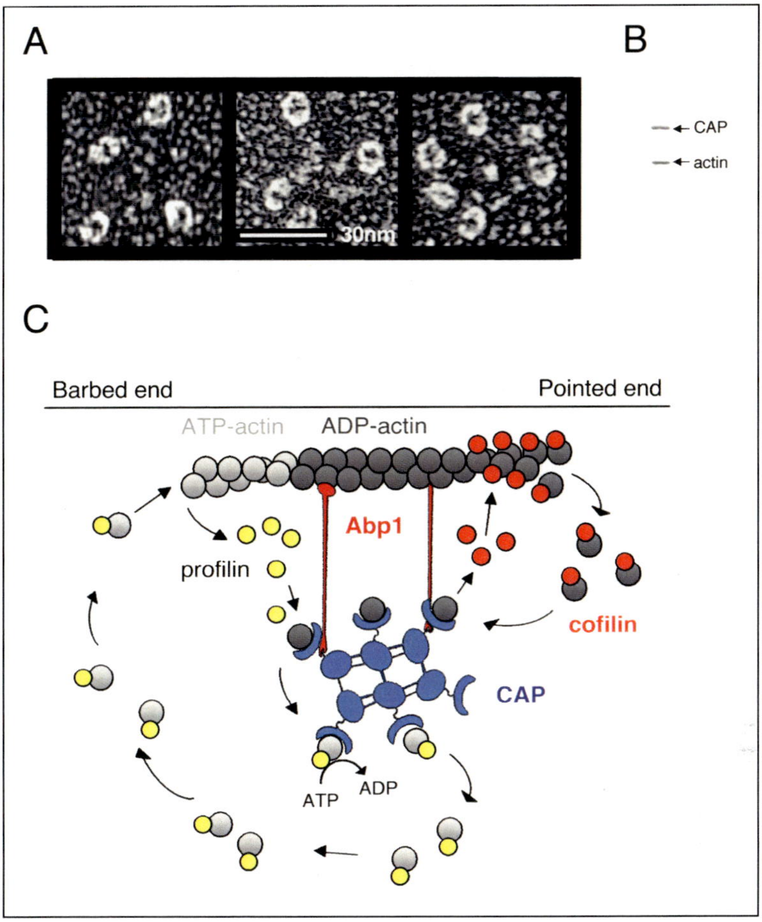

Figure 1. Model for CAP mechanism and function. A) Electron micrographs of purified native *S. cerevisiae* CAP-actin complexes. Samples were adsorbed to mica, then quick-frozen, deep-etched, and rotary-shadowed.[21] B) Coomassie-stained gel of purified CAP-actin complex showing that it consists of only two proteins, CAP (57kDa) and actin (43kDa).[21] C) Model for CAP function in promoting actin turnover (see Mechanism section). Proteins are color-coded: ATP-bound actin subunits, light grey. ADP-bound actin subunits, dark grey. Profilin, yellow. Cofilin, red. CAP, blue.

lobes are depicted as the C-CAP portion of the molecule bound to G-actin. However, further analysis of Srv2 oligomers in the absence of actin is required to test this prediction. Ultimately, elucidating the detailed architecture of the CAP complexes will provide a deeper understanding of the CAP mechanism for promoting actin dynamics. One potential approach for achieving this goal is single particle averaging and 3D reconstruction, coupled to docking the atomic structures for N-CAP, C-CAP and G-actin.

X-ray and/or NMR structures have been reported for two domains of CAP (Fig. 2), the N-terminal α-helical domain (*D. discodeum* residues 1-274),[25-27] and the C-terminal β-sheet actin monomer-binding domain (*S. cerevisiae* residues 369-526).[28] Interestingly, both structures are dimers. NMR analysis suggests that the N-terminal 30-50 residues of CAP are unstructured. However, this region contains heptad repeats and is strongly predicted to form coiled coil structure (CC, Fig. 2).[29] Thus, the N-terminus may become structured under specific conditions such as ligand binding. It is also worth noting that a block of conserved residues is located within this region of CAP and is likely to mediate an important function given its conservation across distant species.

The putative coiled coil domain is followed by a dimer of highly folded α-helix bundles (Fig. 2). Each bundle is composed of six anti-parallel helices, and in the dimer the two monomers interact through helices 3 and 4 from each chain. This generates a four-helix interaction with an interface that has mostly hydrophobic character. The fold of this region bears close resemblance to 14-3-3 proteins, despite sharing only 7% sequence identify. This structural relationship is intriguing given that some 14-3-3 proteins directly regulate cofilin/ADF activity.[30,31] Since the N-terminal half of CAP interacts with cofilin-G-actin complexes,[20] this raises the possibility that N-CAP and 14-3-3 proteins bind cofilin through similar mechanisms.

The C-terminal actin binding domain of CAP forms an unusual V-shaped dimer in which each monomer is comprised of six coils of right-handed β-helix flanked by anti-parallel β-strands (Fig. 2).[28] The last two strands of each monomer cross over each other and interact to form a

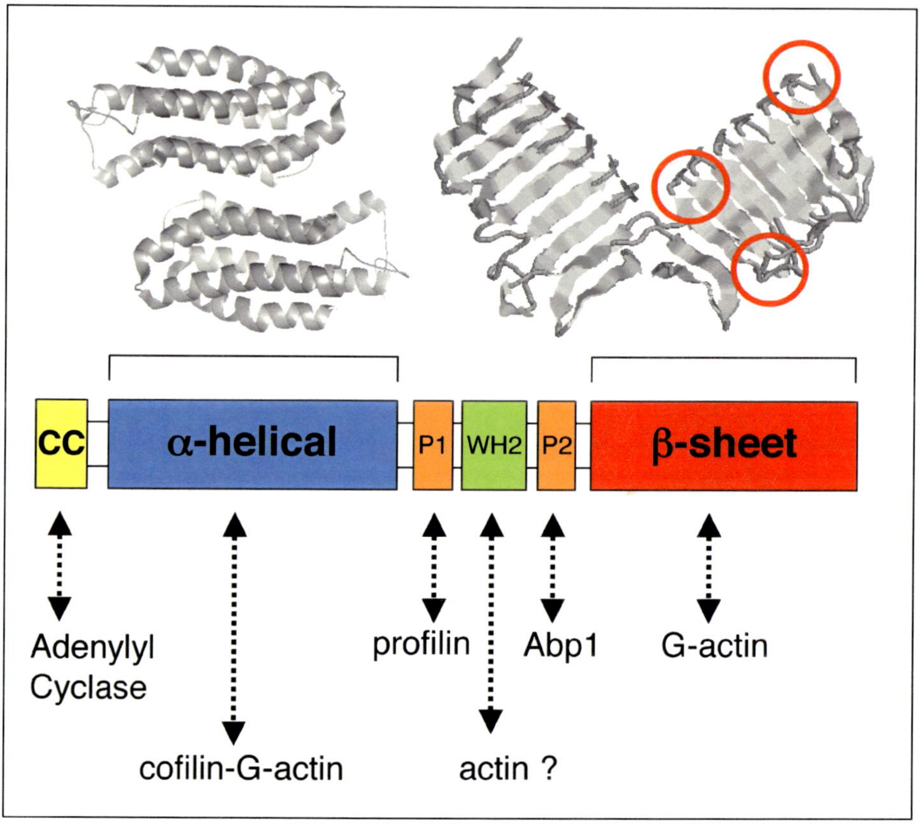

Figure 2. CAP domain organization and atomic structures. CAP domains are colored and labeled: CC, coiled-coil. P1 and P2, proline-rich motifs. WH2, WASp-homology 2. Dotted lines indicate direct physical interactions of specific CAP domains with known ligands. Above the N-terminal α-helical and C-termina l β-sheet domains are their X-ray structures,[28,25] both of which crystallize as dimers. The red circles found on one of the monomers in the β-sheet domain highlight locations of actin binding residues identified by mutational analyses.[22]

"strand-exchanged dimer". In one early study, it was reported that truncation of the C-terminal 27 residues of yeast CAP disrupted G-actin binding.[32] It is now clear that this truncation abolished contacts between monomers and disrupted dimerization. One interpretation of these results is that efficient actin binding requires dimerization of C-CAP.

The structures for N-CAP and C-CAP described above must also be considered in the functional context of the 600kDa CAP-actin complex. Models that involve parallel CAP dimers in the complex are supported by the atomic structures. In addition, the structure of the β-sheet domain puts specific constraints on models for CAP-actin interactions in the complex. Because the two actin binding surfaces of a dimer face in opposite directions, each monomer is likely to make independent binding interactions with G-actin. If these actin binding sites are located at the periphery of the complex, as suggested by the model in Figure 1C, this may facilitate rapid exchange of actin monomers on CAP. It is also important to consider that within the larger complexes, N-CAP may interact with C-CAP, as suggested by two hybrid interaction studies.[32] Still other studies have shown that sequences in the N-terminal coiled coil domain can influence access of ligands to the P2 region.[33] Ultimately, an atomic structure of the intact complex is needed to understand the mechanistic details of CAP function.

Biochemical Properties of the CAP-Actin Association

All CAP proteins examined to date bind actin monomers in a 1:1 molar complex. One interesting property of the interaction is that CAP shows a strong binding preference for ADP-G-actin (Kd = 0.02 μM) compared to ATP-G-actin (Kd = 1.9 μM).[22] This 100-fold difference in binding affinity plays a central role in the CAP mechanism (below). In addition, because CAP binds to ADP-G-actin with about 10-fold higher affinity than cofilin, it competes effectively with cofilin/ADF for binding ADP-G-actin.[22] Although competition between cofilin and CAP suggests that they have partially overlapping binding sites on actin, there are also key differences in their actin-binding interactions. Whereas cofilin/ADF inhibits nucleotide exchange on actin monomers and blocks addition of monomers to the pointed ends of filaments,[1] CAP promotes nucleotide exchange on actin monomers and blocks addition of monomers specifically to the barbed ends of filaments.[20,22]

The actin-binding activity of CAP has been mapped to its C-terminal half. Full actin-binding affinity requires the β-sheet domain and a middle region comprised of two proline-rich motifs (P1 and P2) and a WH2-like domain (Fig. 2). A shorter fragment containing the β-sheet domain alone is sufficient to bind G-actin, but has almost 20-fold lower binding affinity than the longer C-terminal fragment.[22] It remains uncertain how the P1-WH2-P2 region contributes to actin binding affinity since a P1-WH2-P2 fragment has no detectable binding to actin. Mutational analyses guided by the crystal structure have identified three distinct actin-binding surfaces on the β-sheet domain.[22] All three sites map to one side of the flattened structure and are defined by evolutionarily conserved residues. Mutations at each site weaken ADP-actin binding by 4-8 fold and disrupt CAP function in vivo. Thus, high affinity interactions with ADP-G-actin are required for CAP cellular function.

Despite these breakthroughs in CAP structure and function, our understanding of the CAP-actin interaction remains limited. The CAP binding surface on actin has not been defined, and attempts to cocrystallize C-CAP with actin have been unsuccessful (S. Almo and P. Lappalainen, personal communications). Two hybrid studies using a large collection of mutant actin alleles suggest that CAP may contact all four subdomains and both sides of actin (front and back).[34] However, it is difficult to imagine how the CAP β-sheet flattened structure could make such comprehensive binding coverage on actin. Moreover, CAP-G-actin complexes are blocked from being added to the barbed ends of filaments, yet are added readily to the pointed ends of filaments.[22] This suggests that CAP leaves one end of G-actin mostly free, the opposite end from profilin since profilin blocks addition specifically to pointed ends of filaments. Since CAP competes for G-actin binding with both cofilin[22] and gelsolin S1 fragment,[28] the CAP binding site on actin may overlap partially with the footprints of these actin binding proteins.

Clearly, greater efforts are needed to define more precisely the CAP binding site on actin, which will bring mechanistic clarity to CAP function.

There are also lingering questions about the middle region of CAP, which represents the 20% of primary sequence for which there is no atomic structure yet. This region includes proline-rich binding sites for profilin (P1) and the SH3 domain of Abp1 (P2) and a WH2-like domain. The conservation of the WH2-like domain in diverse CAP homologues suggests that it has an important function. Although the majority of WH2 domains bind G-actin,[35] a purified CAP P1-WH2-P2 fragment exhibits no detectable binding to G-actin. Further, a point mutation in the WH2 domain that disrupts actin-binding in other WH2 domains has only a minimal effect on binding to G-actin.[22] These observations raise the possibility that the WH2 domain in CAP has unconventional properties. Possible functions for this domain include binding specifically to nucleotide-free G-actin, binding to F-actin as shown for the WH2 domain in Pan1,[36] or performing a nonactin-binding (possibly structural) role.

CAP Mechanism and Cellular Function

Recent mechanistic studies on yeast and human CAP have drastically revised models for CAP mechanism and function.[20,22] As shown in Figure 1, cofilin severs and depolymerizes actin filaments, which leads to the rapid accumulation of cofilin-bound ADP-actin monomers. These monomers must be recharged to an ATP-bound state for actin to assembly rapidly at free barbed ends. However, cofilin binding blocks nucleotide exchange (ATP for ADP) on G-actin. Profilin has much lower binding affinity for ADP-actin than cofilin, making it inefficient at displacing cofilin from ADP-actin monomers. However, full-length CAP potently catalyzes the displacement of cofilin from ADP-actin monomers.[21] This step may involve the C-terminal actin-binding domain of CAP, which competes strongly with cofilin for binding ADP-G-actin,[22] and possibly the N-terminus of CAP, which binds to cofilin-G-actin complexes.[20] This activity elevates cofilin levels available for filament depolymerization, thereby promoting actin turnover.

The next step in the mechanism is rapid nucleotide exchange on G-actin. This reaction is likely to occur while ADP-G-actin is bound to CAP, because of their high affinity interaction (Kd = 0.02 μM).[22] Profilin may assist CAP in this reaction, as suggested by direct interactions between profilin and the P1 region of CAP.[37,38] Further, profilin appears to be capable of forming a ternary complex with CAP and G-actin (Bertling et al, submitted). Because profilin has strong binding affinity for ATP-actin and CAP has weak affinity for ATP-actin, after nucleotide exchange has occurred on actin profilin-bound ATP-actin complexes dissociate from CAP. This replenishes the cellular pool of assembly-competent actin monomers available for actin polymerization at free barbed ends. Thus, by two distinct, yet coupled mechanisms (recycling of cofilin and processing of actin monomers), CAP promotes the rapid turnover of actin networks.

The mechanism described above is consistent with all available biochemical data for CAP and is supported by multiple lines of genetic data. Early studies showed that profilin overexpression partially suppresses *cap* mutant defects in cell growth and actin organization[8] and profilin is upregulated upon loss of CAP.[10] Further, yeast *cap* mutants show synthetic lethal interactions with *pfy1* alleles, *cof1* alleles, and regulators of cofilin such as *aip1*.[21]

In Vivo Regulation of CAP

Although CAP is abundant in cells, present in yeast at about a 1:10 ratio to actin,[21] surprisingly little is known about its regulation. Post-translational modifications have not been reported for CAP, and little is known about how its protein binding partners influence its activity. One of the few forms of regulation reported for CAP is inhibition of actin binding by phosphatidylinositol 4,5-bisphosphate (PI[4,5]P_2). Interestingly, PI[4,5]P_2 inhibits actin binding of full-length *Dictyostelium* CAP but not a C-CAP fragment,[10] suggesting a possible long-range mechanism for its regulation. Inhibition of CAP-actin interactions by PI(4,5)P_2, thereby decreasing cofilin-dependent actin turnover, is consistent with the known role of PI(4,5)P_2 in down-regulating actin filament disassembly in cells.

Another important aspect of CAP regulation is its localization to filamentous actin networks. As mentioned above, all plant, animal and fungal CAP homologues localize to actin filament structures in cells. In yeast, CAP localization to cortical actin patches relies on a direct interaction between its second proline-rich motif (P2) and the SH3 domain of Abp1[33,39] and purified Abp1 is sufficient to recruit CAP to actin filaments in vitro.[21] The potential effects of Abp1 binding on CAP activity have not been explored. Since P2 is not conserved in vertebrate CAPs, it is unclear how CAP is recruited to actin structures in those cells. In *Drosophila* CAP, the P2 region is hypothesized to interact with Abl tyrosine kinase in the regulation of axon guidance signaling.[12,40] Thus, it is possible that Abl and/or other SH3 domain-containing protein recruit CAP to actin structures.

As mentioned above, CAP activity is important for endocytosis, cell morphogenesis, and cell motility. In addition, CAP may have other functions. For instance, a possible role for CAP in secretion has been suggested by the genetic observation that yeast *cap* mutants are suppressed by overexpression of *SNC1/SNC2*. These yeast genes encode homologues of synaptic vesicle membrane proteins that are known to facilitate vesicle docking and fusion with the plasma membrane.[41,42] Clearly, there is much to learn about the diversity of physiological processes involving CAP. Further, the molecular mechanisms controlling CAP localization and activity in cells have barely begun to emerge.

References

1. Bamburg JR. Proteins of the ADF/cofilin family: Essential regulators of actin dynamics. Annu Rev Cell Dev Biol 1999; 15:185-230.
2. Huang TY, DerMardirossian C, Bokoch GM. Cofilin phosphatases and regulation of actin dynamics. Curr Opin Cell Biol 2006; 18(1):26-31.
3. Ono S. Regulation of actin filament dynamics by actin depolymerizing factor/cofilin and actin-interacting protein 1: New blades for twisted filaments. Biochemistry 2003; 42(46):13363-13370.
4. Fedor-Chaiken M, Deschenes RJ, Broach JR. SRV2, a gene required for RAS activation of adenylate cyclase in yeast. Cell 1990; 61(2):329-340.
5. Field J, Nikawa J, Broek D et al. Purification of a RAS-responsive adenylyl cyclase complex from Saccharomyces cerevisiae by use of an epitope addition method. Mol Cell Biol 1988; 8(5):2159-2165.
6. Field J, Xu HP, Michaeli T et al. Mutations of the adenylyl cyclase gene that block RAS function in Saccharomyces cerevisiae. Science 1990; 247(4941):464-467.
7. Gerst JE, Ferguson K, Vojtek A et al. CAP is a bifunctional component of the Saccharomyces cerevisiae adenylyl cyclase complex. Mol Cell Biol 1991; 11(3):1248-1257.
8. Vojtek A, Haarer B, Field J et al. Evidence for a functional link between profilin and CAP in the yeast S. cerevisiae. Cell 1991; 66(3):497-505.
9. Hubberstey AV, Mottillo EP. Cyclase-associated proteins: Capacity for linking signal transduction and actin polymerization. FASEB J 2002; 16(6):487-499.
10. Gottwald U, Brokamp R, Karakesisoglou I et al. Identification of a cyclase-associated protein (CAP) homologue in Dictyostelium discoideum and characterization of its interaction with actin. Mol Biol Cell 1996; 7(2):261-272.
11. Noegel AA, Rivero F, Albrecht R et al. Assessing the role of the ASP56/CAP homologue of Dictyostelium discoideum and the requirements for subcellular localization. J Cell Sci 1999; 112(Pt 19):3195-3203.
12. Baum B, Li W, Perrimon N. A cyclase-associated protein regulates actin and cell polarity during Drosophila oogenesis and in yeast. Curr Biol 2000; 10(16):964-973.
13. Benlali A, Draskovic I, Hazelett DJ et al. act up controls actin polymerization to alter cell shape and restrict Hedgehog signaling in the Drosophila eye disc. Cell 2000; 101(3):271-281.
14. Barrero RA, Umeda M, Yamamura S et al. Arabidopsis CAP regulates the actin cytoskeleton necessary for plant cell elongation and division. Plant Cell 2002; 14(1):149-163.
15. Bertling E, Hotulainen P, Mattila PK et al. Cyclase-associated protein 1 (CAP1) promotes cofilin-induced actin dynamics in mammalian nonmuscle cells. Mol Biol Cell 2004; 15(5):2324-2334.
16. Zelicof A, Gatica J, Gerst JE. Molecular cloning and characterization of a rat homolog of CAP, the adenylyl cyclase-associated protein from Saccharomyces cerevisiae. J Biol Chem 1993; 268(18):13448-13453.

17. Kawamukai M, Gerst J, Field J et al. Genetic and biochemical analysis of the adenylyl cyclase-associated protein, cap, in Schizosaccharomyces pombe. Mol Biol Cell 1992; 3(2):167-180.
18. Gieselmann R, Mann K. ASP-56, a new actin sequestering protein from pig platelets with homology to CAP, an adenylate cyclase-associated protein from yeast. FEBS Lett 1992; 298(2-3):149-153.
19. Freeman NL, Chen Z, Horenstein J et al. An actin monomer binding activity localizes to the carboxyl-terminal half of the Saccharomyces cerevisiae cyclase-associated protein. J Biol Chem 1995; 270(10):5680-5685.
20. Moriyama K, Yahara I. Human CAP1 is a key factor in the recycling of cofilin and actin for rapid actin turnover. J Cell Sci 2002; 115(Pt 8):1591-1601.
21. Balcer HI, Goodman AL, Rodal AA et al. Coordinated regulation of actin filament turnover by a high-molecular-weight Srv2/CAP complex, cofilin, profilin, and Aip1. Curr Biol 2003; 13(24):2159-2169.
22. Mattila PK, Quintero-Monzon O, Kugler J et al. A high-affinity interaction with ADP-actin monomers underlies the mechanism and in vivo function of Srv2/cyclase-associated protein. Mol Biol Cell 2004; 15(11):5158-5171.
23. Wang J, Suzuki N, Kataoka T. The 70-kilodalton adenylyl cyclase-associated protein is not essential for interaction of Saccharomyces cerevisiae adenylyl cyclase with RAS proteins. Mol Cell Biol 1992; 12(11):4937-4945.
24. Yang S, Cope MJ, Drubin DG. Sla2p is associated with the yeast cortical actin cytoskeleton via redundant localization signals. Mol Biol Cell 1999; 10(7):2265-2283.
25. Ksiazek D, Brandstetter H, Israel L et al. Structure of the N-terminal domain of the adenylyl cyclase-associated protein (CAP) from Dictyostelium discoideum. Structure 2003; 11(9):1171-1178.
26. Mavoungou C, Israel L, Rehm T et al. NMR structural characterization of the N-terminal domain of the adenylyl cyclase-associated protein (CAP) from Dictyostelium discoideum. J Biomol NMR 2004; 29(1):73-84.
27. Yusof AM, Hu NJ, Wlodawer A et al. Structural evidence for variable oligomerization of the N-terminal domain of cyclase-associated protein (CAP). Proteins 2005; 58(2):255-262.
28. Dodatko T, Fedorov AA, Grynberg M et al. Crystal structure of the actin binding domain of the cyclase-associated protein. Biochemistry 2004; 43(33):10628-10641.
29. Nishida Y, Shima F, Sen H et al. Coiled-coil interaction of N-terminal 36 residues of cyclase-associated protein with adenylyl cyclase is sufficient for its function in Saccharomyces cerevisiae ras pathway. J Biol Chem 1998; 273(43):28019-28024.
30. Gohla A, Bokoch GM. 14-3-3 regulates actin dynamics by stabilizing phosphorylated cofilin. Curr Biol 2002; 12(19):1704-1710.
31. Birkenfeld J, Betz H, Roth D. Identification of cofilin and LIM-domain-containing protein kinase 1 as novel interaction partners of 14-3-3 zeta. Biochem J 2003; 369(Pt 1):45-54.
32. Zelicof A, Protopopov V, David D et al. Two separate functions are encoded by the carboxyl-terminal domains of the yeast cyclase-associated protein and its mammalian homologs. Dimerization and actin binding. J Biol Chem 1996; 271(30):18243-18252.
33. Yu J, Wang C, Palmieri SJ et al. A cytoskeletal localizing domain in the cyclase-associated protein, CAP/Srv2p, regulates access to a distant SH3-binding site. J Biol Chem 1999; 274(28):19985-19991.
34. Amberg DC, Basart E, Botstein D. Defining protein interactions with yeast actin in vivo. Nat Struct Biol 1995; 2(1):28-35.
35. Paunola E, Mattila PK, Lappalainen P. WH2 domain: A small, versatile adapter for actin monomers. FEBS Lett 2002; 513(1):92-97.
36. Toshima J, Toshima JY, Martin AC et al. Phosphoregulation of Arp2/3-dependent actin assembly during receptor-mediated endocytosis. Nat Cell Biol 2005; 7(3):246-254.
37. Lambrechts A, Verschelde JL, Jonckheere V et al. The mammalian profilin isoforms display complementary affinities for PIP2 and proline-rich sequences. EMBO J 1997; 16(3):484-494.
38. Drees BL, Sundin B, Brazeau E et al. A protein interaction map for cell polarity development. J Cell Biol 2001; 154(3):549-571.
39. Lila T, Drubin DG. Evidence for physical and functional interactions among two Saccharomyces cerevisiae SH3 domain proteins, an adenylyl cyclase-associated protein and the actin cytoskeleton. Mol Biol Cell 1997; 8(2):367-385.
40. Baum B, Perrimon N. Spatial control of the actin cytoskeleton in Drosophila epithelial cells. Nat Cell Biol 2001; 3(10):883-890.
41. Protopopov V, Govindan B, Novick P et al. Homologs of the synaptobrevin/VAMP family of synaptic vesicle proteins function on the late secretory pathway in S. cerevisiae. Cell 1993; 74(5):855-861.
42. Couve A, Gerst JE. Yeast Snc proteins complex with Sec9. Functional interactions between putative SNARE proteins. J Biol Chem 1994; 269(38):23391-23394.

Chapter 5

Twinfilin Family of Actin Monomer-Binding Proteins

Elisa M. Nevalainen, Ville O. Paavilainen and Pekka Lappalainen*

Abstract

Twinfilin family actin monomer-binding proteins are conserved in evolution from yeasts to mammals. They bind ADP-actin monomers with high affinity and prevent the assembly of actin monomers into filament ends. In addition to monomeric actin, twinfilins also bind and cap actin filament barbed ends, and interact directly with heterodimeric capping proteins. Interaction with capping protein is necessary for twinfilin's localization to the cortical actin cytoskeleton at least in budding yeast. Genetic studies on yeast and *Drosophila* demonstrate that twinfilin is intimately involved in the regulation of actin dynamics in cells, and that the lack of twinfilin results in uncontrolled actin filament assembly. Together, these data suggest that twinfilins play an important role in actin dynamics by preventing unwanted actin filament assembly in cells. However, the exact mechanism by which twinfilin regulates actin filament turnover and contributes to actin-dependent cellular processes remains to be elucidated.

Introduction

The actin cytoskeleton plays an essential role in a wide variety of cellular functions involving membrane dynamics. These include e.g., cell division, polarized growth, morphogenesis, motility, and endocytosis. In cells, actin exists in both monomeric (G-actin) and filamentous (F-actin) forms. The dynamics and organization of the actin cytoskeleton are both spatially and temporally regulated by a large array of actin-binding proteins,[1,2] many of which interact with filamentous actin. In addition to regulating the organization of actin filament structures, F-actin binding proteins can also nucleate the formation of new filaments (e.g., Arp2/3 complex, formins, and Spir), prevent the assembly of monomers into filament ends (e.g., gelsolin, capping protein) or promote filament depolymerization (ADF/cofilin family proteins).[3-8] Although actin filaments are considered as the functional state of actin, also actin monomer-binding proteins play an important role in cytoskeletal dynamics by regulating the size, localization and the nucleotide status of the cellular actin monomer pool, and thus controlling the assembly of actin monomers into filament ends.[9]

Despite the large variation in the biochemical activities of actin-binding proteins, many of them interact with actin through a relatively small number of actin-binding motifs. The actin-depolymerizing-factor homology (ADF-H) domain is a structurally conserved ~150-amino acid motif that is capable of interacting with either actin monomers, actin filaments or both.[10] The ADF-H domain is present in five phylogenetically distinct classes of actin-binding proteins: ADF/cofilins, Abp1/drebrins, coactosins, GMFs (glia maturation factors) and twinfilins

*Corresponding Author: Pekka Lappalainen—Institute of Biotechnology, P.O.Box 56, 00014 University of Helsinki, Finland. Email: pekka.lappalainen@helsinki.fi

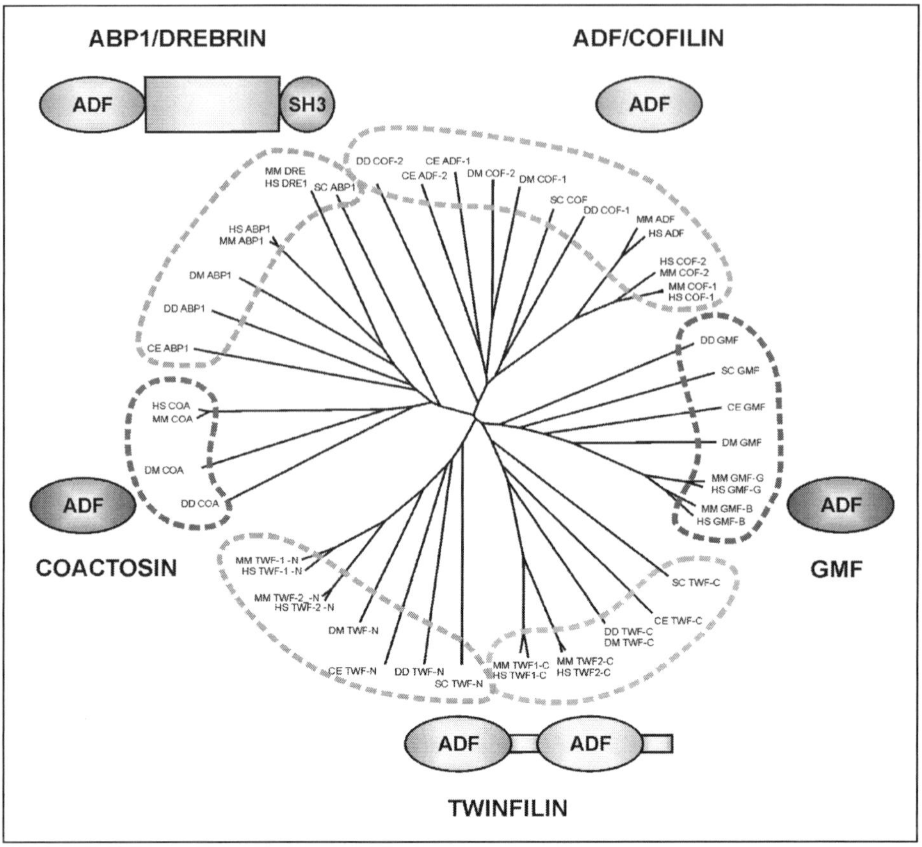

Figure 1. An unrooted phylogenetic tree, prepared from a structure-based sequence alignment of human (HS), mouse (MM), *C.elegans* (CE), *D. melanogaster* (DM), *D. discoideum* (DD), and *S. cerevisiae* (SC) ADF-H domains. ADF/cofilin, coactosin and GMF are composed of single ADF-H domains that are ~20 % identical to each other at the amino acid level. Twinfilin is composed of two ADF-H domains, whereas Abp1/drebrins contain an N-terminal ADF-H domain and a C-terminal SH3 domain. It is important to note that the five functionally distinct classes of ADF-H domain proteins form separate branches in the phylogenetic tree. Also the N- and C-terminal ADF-H domains of twinfilin form independent groups, suggesting that the duplication of the ADF-H domain was an ancient event that took place before the divergence of yeast and animal lineages.

(Fig. 1). Although all these proteins use the ADF-H domain for their interactions with actin, they are biochemically distinct and play different roles in actin dynamics. ADF/cofilins, coactosins, and Abp1/drebrins are F-actin-binding proteins, while twinfilins bind actin monomers and filament barbed ends.[8,11-14] The possible actin interactions of GMFs have not been characterized so far. In this review, we will discuss the function of twinfilin, a ubiquitous actin monomer-binding protein that is composed of two ADF-H domains.

Structural Features of Twinfilin

Twinfilin was first identified from a human cell line as a novel tyrosine kinase and consequently named 'A6 protein tyrosine kinase'.[15] However, twinfilin does not contain any of the sequence motifs commonly conserved in the catalytic domains of protein kinases, and subsequent studies on yeast and mouse proteins demonstrated that twinfilins do not display any kinase activity, but instead function as actin monomer-binding proteins.[16,17]

Genes encoding twinfilins have so far been characterized from *S. cerevisiae, S. pombe, C. elegans, Drosophila,* mouse, and human, suggesting that twinfilins are present in eukaryotes from yeast to mammals. However, twinfilins appear not to be present in plants.[10,13,16-18] All twinfilins are composed of two ADF-H domains arranged in tandem, separated by a short linker region, and followed by a flexible C-terminal tail region.[13] The two ADF-H domains of twinfilin show ~20% amino acid sequence identity to other ADF-H domain proteins and ~25% identity to each other (Fig. 1). The individual ADF-H domains of twinfilin are more closely related across species than they are to the other ADF-H domain within a given protein, so the two domains form separate branches in a phylogenetic tree (Fig. 1). Thus, the duplication of an ADF-H domain was an ancient event that took place already before the divergence of yeast and animal lineages.

The atomic structure of twinfilin's N-terminal ADF-H domain has been determined by X-ray crystallography.[19] It has very similar overall fold to ADF/cofilins, coactosins, and the ADF-H domain of Abp1, demonstrating that the ADF-H domains form a structurally conserved actin-binding motif. The actin monomer-binding interfaces are chemically very similar between ADF/cofilins and twinfilin's N-terminal domain, and the largest differences are seen in regions that are important for actin filament-binding in ADF/cofilin. These differences provide a structural explanation for the variations on F-actin-binding properties between these two proteins.[19]

Biochemical Properties and Interaction Partners of Twinfilin

All twinfilins from yeast to mammals bind actin monomers.[16-18,20] Twinfilin, although composed of two potential actin-binding motifs (one in each ADF-H domain), forms a stable, high-affinity 1:1 complex with G-actin.[16,17,20] Upon binding twinfilin inhibits the nucleotide exchange of actin monomers and prevents the assembly of these monomers into actin filaments.[16,21]

Twinfilin interacts with ADP-G-actin with ~10-fold higher affinity than with ATP-G-actin. Despite the functional differences between cofilin and twinfilin, these proteins compete in actin monomer binding and probably interact with actin monomers through at least partially overlapping sites.[20] The affinity of twinfilin for ADP-G-actin is slightly higher than that of ADF/cofilins.[22-25] This, together with a relatively high cellular concentration of twinfilin, suggests that twinfilin might displace ADF/cofilin from ADP-actin in vivo.[20]

The actin-binding activity of twinfilin resides entirely in the two ADF-H domains; the linker and tail regions are not involved.[20] In yeast twinfilin, the presence of both ADF-H domains is required for efficient G-actin binding.[16,26] In contrast, mouse twinfilin ADF-H domains are able to bind actin monomers individually. The strong actin monomer-binding and -sequestering activities are located in the C-terminal ADF-H domain, which has approximately 10-fold higher affinity for ADP-G-actin than the N-terminal domain. Kinetic analyses suggest that when twinfilin binds G-actin, the monomer first associates with the N-terminal ADF-H domain, the binding induces a conformational change in the twinfilin molecule, and the actin monomer is then delivered to the C-terminal, high-affinity, ADF-H domain.[20]

Two recent studies showed that twinfilins can also interact with actin filaments. Mouse twinfilin is able to bind to actin filament's barbed end and prevent the assembly of actin monomers into these filament ends. Twinfilin caps barbed ends with higher affinity when terminal F-actin subunits have ADP bound than when ATP or ADP-Pi is bound. Both ADF-H domains are required for the capping activity, and thus this result may explain why twinfilin is composed of two ADF-H domains. Furthermore, biomimetic motility assays demonstrated that twinfilin can replace, through its barbed end binding activity, capping proteins in actin-based motility.[14] Studies on yeast twinfilin demonstrated that it promotes actin filament severing in a pH-dependent manner. Yeast twinfilin functions as an actin monomer sequestering protein at neutral pH, but at acidic environment (pH <6.0) it gains actin filament-binding and severing activities.[27] However, the cellular roles of twinfilin's barbed end capping and F-actin severing activities remain to be shown.

In addition to actin monomers and filament barbed ends, twinfilins also bind heterodimeric capping proteins. These evolutionarily conserved proteins are composed of two subunits and interact with actin filament barbed ends to prevent filament polymerization.[7] Twinfilin - capping protein interaction is conserved in evolution from yeast to mammals.[21,26] The capping protein-binding site is located at the conserved C-terminal tail region of twinfilin and it does not overlap with the actin-binding site of twinfilin. At least in budding yeast, twinfilin's ability to bind to both actin monomers and capping protein appears to be necessary for its role in actin dynamics. Direct interaction between twinfilin and capping protein does not affect their actin monomer sequestering and filament barbed end capping activities.[28] However, capping protein directly inhibits yeast twinfilin binding to and severing of actin filaments at least in vitro.[27]

Phenotypes and Cell Biological Functions

Twinfilin is an abundant protein in yeast, found in a 1:10 ratio to actin and 1:2.5 ratio to cofilin.[26] In yeast cells, twinfilin shows diffuse cytoplasmic staining, but also localizes strongly to the cortical actin patches.[16,26] The ability to interact with capping protein is essential for twinfilin's subcellular localization to the cortical actin cytoskeleton at least in budding and fission yeasts.[26,28,29]

In budding yeast, deletion of twinfilin gene does not cause a severe phenotype, but results in slightly enlarged cortical actin patches, suggesting that twinfilin may limit actin filament assembly in vivo. However, in combination with a temperature-sensitive cofilin mutant *cof1-22*, shown to cause pronounced reduction in turnover of cortical actin filaments,[30] twinfilin causes lethality at the permissive temperature. Synthetic lethality is observed also when *twf*Δ mutation is combined with profilin mutant *pfy1-4*. Profilin is a ubiquitous actin monomer-binding protein, and in *pfy1-4* cells the exchange of actin nucleotide form ADP to ATP is defective due to mutations in profilin gene.[16,31] Interestingly, also twinfilin point mutations that affect its ability to interact with actin monomers or capping protein display synthetic lethality with these cofilin and profilin mutations.[28] These data show that twinfilin is intimately involved in regulation of actin dynamics in yeast cells.

In *Drosophila* twinfilin is ubiquitously expressed at different tissues and developmental stages.[18] A strong hypomorphic mutation in the *twf* gene leads to a number of developmental defects. The mutant flies are slightly smaller, less active and have reduced flight ability as compared to wild-type flies. Also the hatching frequency is reduced and larval period prolonged in the mutants. The most obvious external phenotype of the twinfilin mutant flies are defects in the bristle morphology (Fig. 2). Adult bristles of *Drosophila* are chitinous structures with ridges and grooves that run along the bristles. The bristle shaft is formed from a single cell, and the external shape of the adult bristle is determined by actin filament bundles dispersed at the plasma membrane along the length of the developing bristle. In wild type flies twinfilin is highly abundant in all hair- and bristle-producing cells. Twinfilin localizes diffusively to the cytoplasm and to cortical actin filament structures throughout the developing bristle shaft. In the *twf* mutant flies, the adult bristles are often split, branched, or bent and have a highly irregular ridge pattern. This aberrant bristle morphology in *twf* mutant flies is caused by uncontrolled polymerization of actin filaments resulting in twisted and misoriented actin bundles in developing bristles.[18]

While other organisms, like yeasts and *Drosophila*, have only one twinfilin protein, mammals have two twinfilin isoforms, twinfilin-1 and twinfilin-2.[21] Both isoforms bind ADP-G-actin with high affinity, inhibit the nucleotide exchange on actin monomers, and efficiently prevent actin filament assembly. Both isoforms directly interact with capping protein and with PI(4,5)P_2. Although biochemically very similar, mouse twinfilin isoforms have

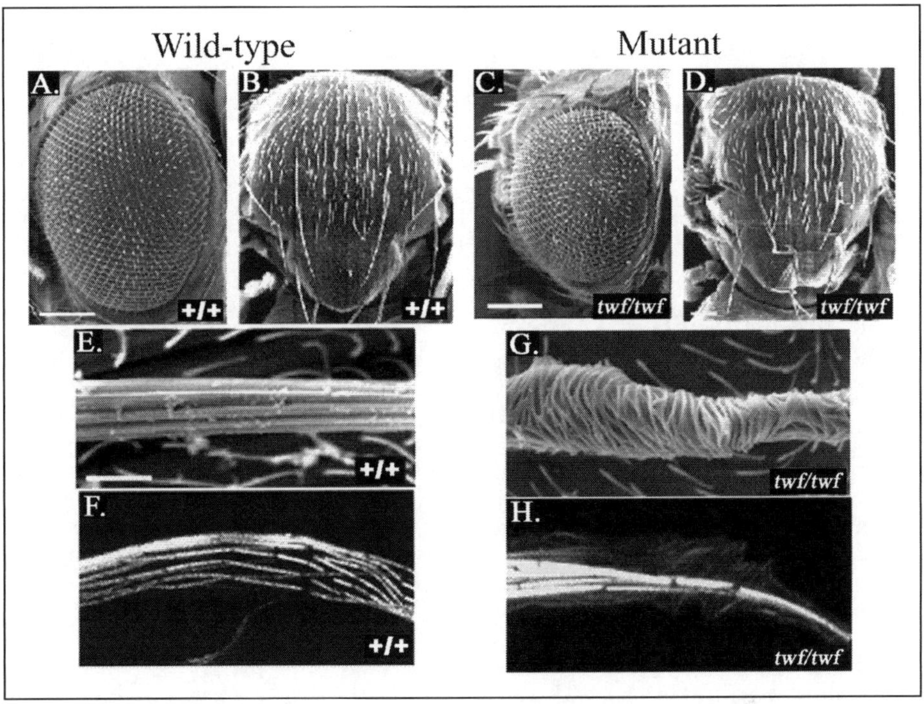

Figure 2. *Drosophila twinfilin* mutants have defects in bristle morphology and in eyes. Scanning electron micrographs of bristles and eyes (A-E,G) and projections of confocal sections through pupal bristles (F,H). A) Wild-type eye and (C) *twf* mutant eye. In *twf* mutants the interommatidial bristles are often tufted and the ommatidia are occasionally fused or pitted. B) The whole thorax from wild-type and (D) *twf* mutant adult. Note the bent and split bristles. E) Surface of a wild-type macrochaeta showing straight longitudinal ridges. F) Middle portion of a wild-type bristle showing straight actin bundles. G) Thickening on a mutant *twf* macrochaeta with ridges arranged perpendicularly to the long axis of the bristle. H) In *twf* mutant actin bundles are misoriented perpendicular to the long axis and form a knob on the bristle. Bars: (A-D) 100 um; (E-H) 10 um.

distinct expression patterns. Either twinfilin-1 or twinfilin-2 is expressed in virtually all cell types of mouse embryos and adult mice. Twinfilin-1 is the major isoform during development and is expressed with variable levels in most adult mouse nonmuscle cell-types. Twinfilin-2 expression is relatively weak during embryonic stages, but it is the predominant isoform of adult heart and skeletal muscles. Both mouse twinfilins are especially abundant in the mechanosensory hair cells of the inner ear. These cells have actin-based apical projections called stereocilia that display structural and functional similarity to *Drosophila* mechanosensory bristles. High levels of twinfilins in mammalian inner ear hair cells may thus be necessary to maintain proper stereocilia structure and function.[21]

Twinfilin-1 and twinfilin-2 are abundant proteins in cultured mammalian cells and have similar subcellular localizations in unstimulated cells. Both proteins show punctate cytoplasmic localization that is especially prominent around the nucleus but are also localized to the actin filament-rich filopodia. These structures are also rich in G-actin. The subcellular localizations of the mammalian twinfilin isoforms are regulated by different signalling pathways since only the localization of twinfilin-1 appears to be affected by small GTPases Rac1 and Cdc42.[21] Rac1 and Cdc42 are central regulators of polarized growth and motility in nonmuscle cells,[32] suggesting that twinfilin-1 may be involved in these cellular functions. Strong expression in developing neurons, skin, and olfactory sensory epithelium further support the role of twinfilin-1 in polarized growth. Since the predominant isoform in muscle cells, twinfilin-2, is not regulated by these small GTPases it may be involved in the maintenance of sarcomere structure and perhaps does not promote any Rac1 and Cdc42-induced processes.[21]

The role of twinfilin in mammals has not been examined by genetic methods so far. However, overexpression and RNAi studies demonstrated that twinfilin is important in clathrin mediated endocytosis in mammalian cells.[14,33] In cultured mouse cells endogenous twinfilin colocalizes with actin-rich comet tails of transferrin-positive endosomes, further supporting its role in actin-dependent uptake of endocytic vesicles.[14]

Regulation of Twinfilin's Activity

Given the important role twinfilin has in actin dynamics, it is likely that its activities are tightly regulated. However, the mechanisms of regulation are still largely unclear. Yeast and mouse twinfilin interact with phosphatidylinositol 4,5-bisphosphate ($PI[4,5]P_2$) and the actin monomer-sequestering activity of both twinfilins is downregulated by this interaction.[21,26] $PI(4,5)P_2$ binding also inhibits the filament severing activity of yeast twinfilin.[27] The possible physiological role of twinfilin - $PI(4,5)P_2$ interaction remains to be determined. Because $PI(4,5)P_2$ promotes actin filament assembly and inhibits actin filament disassembly in cells,[34] inhibition of twinfilin by $PI(4,5)P_2$ may serve as a mechanism to reduce actin monomer sequestering at the regions of rapid actin filament nucleation and assembly in cells.

The Role of Twinfilin in Actin Dynamics

Despite the wealth of biochemical data, the exact mechanism by which twinfilin contributes to the regulation of actin dynamics, is currently unknown. Mammalian twinfilins have three actin-related biochemical activities: (1) sequestering of ADP-G-actin, (2) interaction with capping protein, and (3) capping of filament barbed ends (Fig. 3). Yeast twinfilin shares the first two functions, but instead of capping actin filaments it binds and severs them at low pH.[27] All of these functions serve as evidence for twinfilin's role as an important regulator of actin dynamics, and further suggest that twinfilin's role is to prevent actin filament assembly and promote disassembly. Twinfilin sequesters actin monomers dissociated from filament pointed end and keeps them in ADP bound form, thus preventing them from associating into the filament barbed end until needed. Through interaction with capping protein, twinfilin may localize these actin monomers into the sites of rapid actin filament assembly. By capping filament barbed ends twinfilin may prevent filament growth in unwanted regions. Because twinfilin has a higher affinity for filaments with ADP-actin subunits at the end than ATP- or ADP-Pi subunits at the end, twinfilin preferably caps older filaments, or filaments that have been severed. These activities could be especially important in structures where long organized bundles of actin are needed, like in the hair cells of the inner ear, where twinfilin is strongly expressed.

The twinfilin mutation phenotypes in yeast and *Drosophila* result in formation of abnormal actin filament structures and could therefore be a result of uncontrolled actin filament growth. Furthermore, the synthetic phenotypes with profilin and ADF/cofilin mutants in these organisms gives further support for twinfilin's role in preventing undesired actin polymerization in cells.

In the future it will be important to apply genetic, cell biological and biochemical assays to reveal the exact mechanism by which twinfilin, in combination with other actin-binding proteins contributes to actin filament turnover in cells. Furthermore, it will be important to determine the molecular mechanisms by which twinfilin interacts with actin monomers and filament barbed ends, and to elucidate how these interactions are regulated in cells.

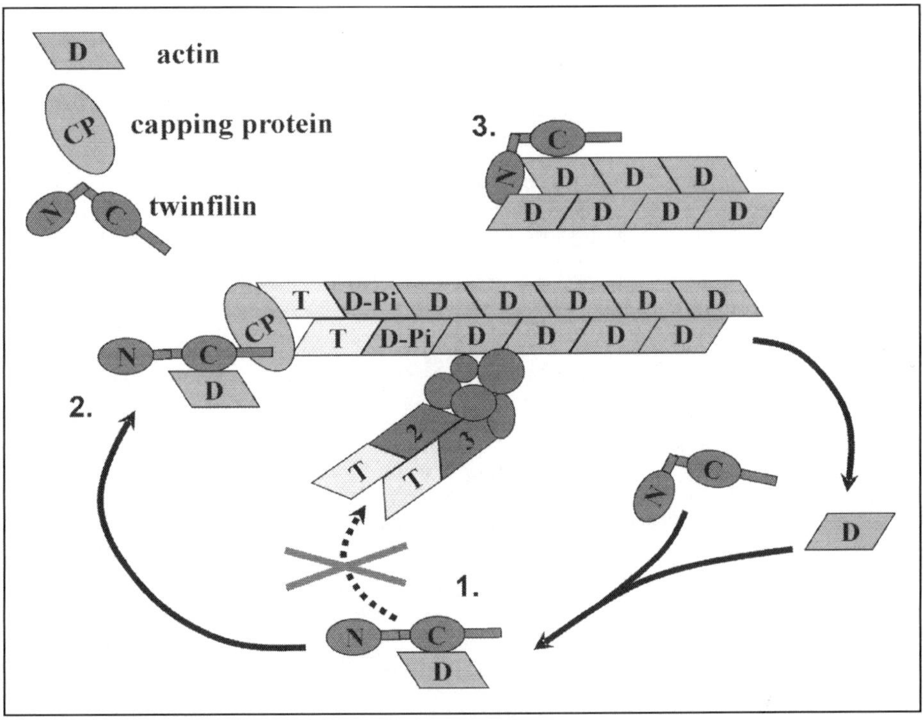

Figure 3. Biochemical activities of twinfilin. All twinfilins examined so far bind ADP-actin monomers with high affinity (1). When bound to an ADP-actin monomer, twinfilin prevents its nucleotide exchange and assembly into filament ends. In addition to actin monomers, twinfilins also interact with capping protein (2). The capping protein-binding site resides at the C-terminal tail-region of twinfilin. Interaction between twinfilin and capping protein does not affect the biochemical activities of these proteins, and thus this interaction may control the localization of twinfilin (or twinfilin-actin monomer complex) to the sites of rapid actin filament assembly in cells. At least mammalian twinfilins also interact with filament barbed ends and prevent the assembly of actin monomers into these filament ends (3).

References

1. Pantaloni D, Le Clainche C, Carlier MF. Mechanism of actin-based motility. Science 2001; 292(5521):1502-6.
2. Pollard TD, Borisy GG. Cellular motility driven by assembly and disassembly of actin filaments. Cell 2003; 112(4):453-65.
3. Welch MD, Mullins RD. Cellular control of actin nucleation. Annu Rev Cell Dev Biol 2002; 18:247-88.
4. Zigmond SH. Formin-induced nucleation of actin filaments. Curr Opin Cell Biol 2004; 16(1):99-105.
5. Quinlan ME, Heuser JE, Kerkhoff E et al. Drosophila Spire is an actin nucleation factor. Nature 2005; 433(7024):382-8
6. McGough AM, Staiger CJ, Min JK et al. The gelsolin family of actin regulatory proteins: Modular structures, versatile functions. FEBS Lett 2003; 552(2-3):75-81.
7. Wear MA, Cooper JA. Capping protein: New insights into mechanism and regulation. Trends Biochem Sci 2004; 29(8):418-28.
8. Bamburg JR, Wiggan OP. ADF/cofilin and actin dynamics in disease. Trends Cell Biol 2002; 12(12):598-605.

9. Paavilainen VO, Bertling E, Falck S et al. Regulation of cytoskeletal dynamics by actin-monomer-binding proteins. Trends Cell Biol 2004; 14(7):386-94.
10. Lappalainen P, Kessels MM, Jamie M et al. The ADF homology (ADF-H) domain: A highly exploited actin-binding module. Mol Biol Cell 1998; 9(8):1951-9.
11. Quintero-Monzon O, Rodal AA, Strokopytov B et al. Structural and functional dissection of the Abp1 ADFH actin-binding domain reveals versatile in vivo adapter functions. Mol Biol Cell 2005; 16(7):3128-39.
12. De Hostos EL, Bradtke B, Lottspeich F et al. Coactosin, a 17 kDa F-actin binding protein from Dictyostelium discoideum. Cell Motil Cytoskeleton 2003; 26(3):181-91.
13. Palmgren S, Vartiainen M, Lappalainen P. Twinfilin, a molecular mailman for actin monomers. J Cell Sci 2002; 115(Pt 5):881-6.
14. Helfer E, Nevalainen EM, Naumanen P et al. Mammalian twinfilin sequesters ADP-G-actin and caps filament barbed ends: Implications on motility. EMBO J 2006; 25(6):1184-95.
15. Beeler JF, LaRochelle WJ, Chedid M et al. Prokaryotic expression cloning of a novel human tyrosine kinase. Mol Cell Biol 1994; 14(2):982-8.
16. Goode BL, Drubin DG, Lappalainen P. Regulation of the cortical actin cytoskeleton in budding yeast by twinfilin, a ubiquitous actin monomer-sequestering protein. J Cell Biol 1998; 142(3):723-33.
17. Vartiainen M, Ojala PJ, Auvinen P et al. Mouse A6/twinfilin is an actin monomer-binding protein that localizes to the regions of rapid actin dynamics. Mol Cell Biol 2000; 20(5):1772-83.
18. Wahlström G, Vartiainen M, Yamamoto L et al. Twinfilin is required for actin-dependent developmental processes in Drosophila. J Cell Biol 2001; 155(5):787-796.
19. Paavilainen VO, Merckel MC, Falck S et al. Structural conservation between the actin monomer-binding sites of twinfilin and actin-depolymerizing factor (ADF)/cofilin. J Biol Chem 2002; 277(45):43089-95.
20. Ojala PJ, Paavilainen VO, Vartiainen MK et al. The two ADF-H domains of twinfilin play functionally distinct roles in interactions with actin monomers. Mol Biol Cell 2002; 13(11):3811-21.
21. Vartiainen MK, Sarkkinen EM, Matilainen T et al. Mammals have two twinfilin isoforms whose subcellular localizations and tissue distributions are differentially regulated. J Biol Chem 2003; 278(36):34347-34355.
22. Blanchoin L, Pollard TD. Interaction of actin monomers with Acanthamoeba actophorin (ADF/cofilin) and profilin. J Biol Chem 1998; 273(39):25106-11.
23. Ressad F, Didry D, Xia GX et al. Kinetic analysis of the interaction of actin-depolymerizing factor (ADF)/cofilin with G- and F-actins. Comparison of plant and human ADFs and effect of phosphorylation. J Biol Chem 1998; 273(33):20894-902.
24. Vartiainen MK, Mustonen T, Mattila PK et al. The three mouse actin-depolymerizing factor/cofilins evolved to fulfill cell-type-specific requirements for actin dynamics. Mol Biol Cell 2002; 13(1):183-94.
25. Yeoh S, Pope B, Mannherz HG et al. Determining the differences in actin binding by human ADF and cofilin. J Mol Biol 2002; 315(4):911-25.
26. Palmgren S, Ojala PJ, Wear MA et al. Interactions with PIP2, ADP-actin monomers, and capping protein regulate the activity and localization of yeast twinfilin. J Cell Biol 2001; 155(2):251-60.
27. Moseley JB, Okada K, Balcer HI et al. Twinfilin severs actin filaments and promotes actin turnover in vivo. J Cell Sci 2006; 119:1547-57.
28. Falck S, Paavilainen VO, Wear MA et al. Biological role and structural mechanism of twinfilin-capping protein interaction. EMBO J 2004; 23(15):3010-9.
29. Kovar DR, Wu JQ, Pollard TD. Profilin-mediated competition between capping protein and formin Cdc12p during cytokinesis in fission yeast. Mol Biol Cell 2005; 16(5):2313-24.
30. Lappalainen P, Drubin DG. Cofilin promotes rapid actin filament turnover in vivo. Nature 1997; 388(6637):78-82.
31. Wolven AK, Belmont LD, Mahoney NM et al. In vivo importance of actin nucleotide exchange catalyzed by profilin. J Cell Biol 2000; 150(4):895-904.
32. Etienne-Manneville S, Hall A. Rho GTPases in cell biology. Nature 2002; 420(6916):629-35.
33. Pelkmans L, Fava E, Grabner H et al. Genome-wide analysis of human kinases in clathrin- and caveolae/raft-mediated endocytosis Nature 2005; 436(7047):78-86.
34. Hilpelä P, Vartiainen MK, Lappalainen P. Regulation of the actin cytoskeleton by PI(4,5)P2 and PI(3,4,5)P3. Curr Top Microbiol Immunol 2004; 282:117-63.

Chapter 6

Intracellular β-Thymosins

Ewald Hannappel,* Thomas Huff and Daniel Safer

Abstract

The β-thymosins are a family of highly conserved polar peptides consisting of 40 to 44 amino acid residues. All β-thymosins bind monomeric G-actin in a 1:1 complex. The dissociation constant of the complex is in the micromolar range and allows for fast binding and release of G-actin. Because of the high intracellular concentration of β-thymosins (up to 500 μM) in most vertebrate cells, β-thymosins are considered the main intracellular G-actin sequestering peptides. Thymosin β_4 binds to G-actin in an extended conformation, and folds into a stable conformation upon binding. The N- and C-termini of thymosin β_4 contact the barbed and pointed ends of the monomeric actin. Thymosin β_4 is present in the nucleus as well as the cytoplasm and might be responsible for sequestering nuclear actin. Even minor cell damage might be responsible for the release of β-thymosins detectable in the extracellular fluids. Extracellular β-thymosins affect matrix metallo-proteinases, chemotaxis, angiogenesis and wound healing. However, only very little is known about the molecular mechanisms mediating the effects attributed to extracellular β-thymosins.

Introduction

Thymosin β_4 was first isolated from extracts of calf thymus as a potential thymic hormone involved in maturation of T-cells and released to the extracellular space.[1] In 1991, it was discovered that thymosin β_4 is identical to Fx, the major intracellular G-actin-sequestering peptide in human platelets.[2] The main topic of this review will be the function of intracellular β-thymosins, for a review on the potential functions of extracellular β-thymosins see reference 3. The function of extra- and intracellular β-thymosins are different and therefore it is of utmost importance to distinguish between data observed by changing the intracellular concentration of β-thymosin and data observed by adding extracellular β-thymosins to cells.

Amino Acid Sequences and Phylogenetic Distribution of β-Thymosins

Currently, 15 β-thymosins from various vertebrates and invertebrates have been isolated and characterized on the peptide level.[4] β-thymosins form a family of highly conserved, extremely polar 5-kDa polypeptides. Members of that family have been found in species ranging from mammals to echinoderms but not in yeast or prokaryotes (Fig. 1). According to a search in EST databases, other species may also contain members of the β-thymosin family. Since none of them has been characterized at the protein level, we will not give the amino acid sequences of those putative peptides. The properties of β-thymosins are rather unique. The presence of an initiator codon immediately preceding the codon for the N-terminal serine or alanine residue and two stop codons following the C-terminal amino acid residue, suggests that β-thymosins are

*Corresponding Author: Ewald Hannappel—Institute of Biochemistry, Faculty of Medicine, University of Erlangen-Nuremberg, Fahrstrasse 17, 91054 Erlangen, Germany. Email: thymosin@biochem.uni-erlangen.de

	1 10 20 30 40 43	
β_4	acSDKPDMAEIEKFDKSKLKKTETQEKNPLPSKETIEQEKQAGES	human, orangutan, rat, mouse, cat, calf, horse, pig, sheep, chicken, gecko
β_4^{Ala}	acADKPDMAEIEKFDKSKLKKTETQEKNPLPSKETIEQEKQAGES	rabbit
β_4^{Xen}	acSDKPDMAEIEKFDKAKLKKTETQEKNPLPSKETIEQEKQSTES	frog
β_9	acADKPDLGEINSFDKAKLKKTETQEKNTLPTKETIEQEKQAK	bovine
β_9^{Met}	acADKPDMGEINSFDKAKLKKTETQEKNTLPTKETIEQEKQAK	pig, sheep
β_{10}	acADKPDMGEIASFDKAKLKKTETQEKNTLPTKETIEQEKRSEIS	human, rat, mouse, cat, calf, horse
β_{11}	acSDKPNLEEVASFDKTKLKKTETQEKNPLPTKETIEQEKQAS	trout
β_{12}	acSDKPDLAEVSNFDKTKLKKTETQEKNPLPTKETIEQEKQATA	trout
β_{12}^{perch}	acSDKPDISEVTSFDKTKLKKTETQEKNPLPSKETIEQEKAAATS	perch
β_{13}	acADKPDMGEIASFDKAKLKKTETQEKNTLPTKETIEQEKQAK	whale
β_{14}	acSDKPDISEVSSFDKTKLKKTETAEKNTLPTKETIEQEKTA	sea urchin
β_{15}	acSDKPDLSEVETFDKSKLKKTNTEEKNTLPSKETIQQEKEYNQRS	human, rat, prostatic adenocarcinoma
$\beta^{scallop}$	acSDKPDVSEVANFDKSKLKKTETAEKNTLPTKETIQQEKSA	scallop
$\beta^{zebrafish}$	acADKPNMTEITSFDKTKLRKTETQEKNPLPTKETIEQERQGESTP	zebrafish
$\beta^{sea\ urchin}$	acADKPDVSEVSTFDKSKLKKTETQEKNTLPTKDTIEQEKQG	sea urchin
	*:***:: *: ***:**:**** *** **:*:**:**:	isolated β-thymosins

Figure 1. Amino acid sequences of β-thymosins from various species. β_4 (P01253); β_4^{Ala} (P34032); β_4^{Xen} (*Xenopus laevis*, P18758); β_9 (P21752); β_9^{Met} (P21753); β_{10} (P13472); β_{11} (*Oncorhynchus mykiss*, P26351); β_{12} (*Oncorhynchus mykiss*, P26352); β_{12}^{perch} (*Lateolabrax japonicus*, P33248); β_{13} (according to ref. 63); β_{14} (according to ref. 64); β_{15} (P97563); $\beta^{scallop}$ (*Argopectan irradians*, according to ref. 16); $\beta^{zebrafish}$ (Q9W7M8); $\beta^{sea\ urchin}$ (*Arbacia punctulata*, according to ref. 16). The numbers in parentheses are the accession codes for the SWISS-PROT databases. Invariant residues are indicated by asterisks and highly conserved residues are represented by colons. β-Thymosins identified only by nucleic acid sequencing are not depicted. For those sequences β_{4y} (O14604), $\beta_4^{neuroblastoma}$ (Q99406), β_{15}^{mouse} (Q9D2R9), β^{carp-A} (*Cyprinus carpio*, Q91955), β^{carp-B} (Q91954), β^{sycon} (*Sycon raphanus*, Q9GUA6), $\beta^{strongyl}$ (*Strongylocentrotus purpuratus*, O76538) and β^{gilli} (*Gillichthys mirabilis*, Q9DFJ) (see TrEMBL databases or ref. 65).

synthesized without formation of a larger precursor.[5] The N-terminal residue is always acetylated. Except for one conserved phenylalanine residue present at position 12 and a tyrosine residue at position 40 of β_{15}, β-thymosins do not contain aromatic amino acid residues. Thus, their absorption at 280 nm is only marginal and they have to be detected by absorption below 220 nm.

Cell and Tissue Distribution

Thymosin β_4 is usually the main β-thymosin peptide in vertebrate tissues, representing about 70-80% of the total β-thymosins in normal tissue of adult rats (450 mg/kg spleen, 11 mg / kg muscle).[6] It is present in high concentrations (up to 400 μM) in rodent tissue, tumor cells and cell lines from various mammalian species.[7,8] EBV-transformed human B cells contain up to 1 pg of thymosin β_4 per cell and 1% of total protein biosynthesis is dedicated to its synthesis.[9] Thymosin β_4 is present in most mammalian tissues and circulating white blood cells, but is detectable in neither human nor avian red blood cells.[4,10] It is the most abundant G-actin-sequestering protein (intracellular concentration about 200 to 500 μM) in human blood platelets and white cells.[10,11] In humans, three β-thymosins have been identified: thymosins β_4 (main peptide), β_{10} (0.05 to 0.3 relative to β_4) and β_{15} (detected in prostate and breast cancer cells and developing neurons).[12-14] The mammalian gene encoding thymosin β_4 is localized on the X chromosome.[15]

Structure of β-Thymosins

NMR studies show that β-thymosins are mostly unstructured in aqueous solution and fold on binding to G-actin.[16-19] The ^{17}LKKTET^{22} sequence conserved in most β-thymosins is considered to be responsible for the initial interaction with G-actin (Fig. 1). Safer et al have already proposed in their initial publication on G-actin sequestering that the sequence motif ^{17}LKKTETQEK^{25} of thymosin β_4 may be involved in actin binding because of its high homology with the well-known actin-binding sequence of actobindin.[2,20,21] Full-length variants of thymosin β_4 generated by chemical synthesis were used to study the requirements for binding and sequestering G-actin. The N-terminal part of thymosin β_4 needs to form an α-helical conformation, the three hydrophobic residues (^{6}M,^{9}I,^{12}F) are important for interaction and the motif ^{17}LKK^{19} are of major importance for interaction.[22] It has also been shown that an extension of the N-terminal α-helical region results in loss of G-actin sequestering activity.[23]

Intracellular Distribution of Thymosin β_4

Thymosin β_4 is localized to both the cytoplasm and the nucleus of cells.[24] Because of its small size, thymosin β_4 might enter the nucleus by diffusion via the nuclear pore. However, free diffusion does not explain why thymosin β_4 is sometimes enriched in the nucleus when fluorescently labeled thymosin β_4 is injected into cells or added to permeabilized cells. It has been recently confirmed that the nucleus contains several other actin-binding proteins and significant amounts of actin not in the typical filamentous form. The function of this protein network is under active investigation.[25,26] Thymosin β_4 might also be the main G-actin sequestering protein in the nucleus. Because considerable amounts of thymosin β_4 are present in the nucleus, it is possible that it might act as a transcription factor.[24]

β-Thymosin/WH2 Domain

The WH2 domain (Wiscott-Aldridge syndrome protein homology domain-2) is an actin monomer-binding motif; the minimal motif was originally proposed to be 18 amino acid residues[27] This domain is found in different regulators of the actin cytoskeleton. Among these proteins are yeast verprolin, human WIP, N-WASP, WAVE and srv2/CAP. In 2002, the boundaries of the WH2 domain were extended to about 35 amino acid residues and it was noticed that β-thymosins (40–44 amino acid residues) are entirely composed of a single WH2 domain.[28] The protein ciboulot expressed in Drosophila contains three repeats of the β-thymosin / WH2 domain and binds to G-actin as thymosin β_4. Unlike conventional β-thymosins ciboulot participates in actin polymerization and is functionally more similar to profilin.[29] The formation of one superfamily (β-thymosin / WH2) from two separate but certainly related families has been disputed.[30,31] Despite low sequence identity, WH2 shares structural similarity with the N-terminal portion of thymosin β_4. However, WH2 is significantly shorter than β-thymosins and binds to actin more strongly. WH2 also lacks the C-terminal region of β-thymosins that becomes involved in monomer sequestration by interfering with subunit contacts in F-actin.[32] There is a further significant difference between the classical WH2 domain containing proteins including those proteins with β-thymosin repeats (ciboulot, tetrathymosin, ...) and classical β-thymosins. While classical WH2 domain containing proteins are folded, β-thymosins are mostly unstructured proteins and fold on binding to G-actin.[17,19] The induction of secondary structure of β-thymosins by interaction with other proteins raises the possibility that β-thymosins might bind various partners and thus perform multiple functions. The term "moonlighting proteins" has been coined to describe this behavior of polypeptides.[33,34] Depending on the location of β-thymosins inside or outside of cells and its secondary structure, β-thymosins might fulfill various biologically important functions in addition to sequestering G-actin.[3]

Sequestering of G-Actin by β-Thymosins

The affinity of β-thymosins for G-actin depends on the specific actin and β-thymosin isoform and is relatively insensitive to ionic conditions but highly sensitive to the nucleotide bound to actin. For thymosin β_4 and human platelet ATP-actin under physiological conditions, the dissociation constant (K_d) is 0.4 to 0.7 μM; for rabbit muscle actin under the same conditions, the K_d is about 2 μM.[11] The affinity of thymosin β_4 for ADP-actin is about 50- to 100-fold lower than for ATP-actin.[35,36] Actin binding affinities have been measured using a variety of methods (change of fluorescence, equilibrium centrifugation, ultrafiltration etc.), for several other β-thymosins. All reported K_d values were in the micromolar range.[12,16,35,37-40]

The stability of the G-actin – β-thymosin complex can be altered by changes in the amino acid sequence of thymosin β_4 or its truncation. This is most impressive in the case of thymosin β_4^{Ala}, where only the first amino acid residue is changed from serine to alanine, which results in a three- to fivefold higher affinity to G-actin.[37] While human tissues contain thymosin β_4 as the major and β_{10} as the minor peptide, human leukocytes contain almost exclusively thymosin β_4. The picture is quite different in rabbit tissues, which contain thymosin β_4^{Ala} and β_{10}. Since the affinities of thymosin β_4 and β_{10} for G-actin are almost identical while thymosin β_4^{Ala} has a higher affinity for G-actin, this might explain why in rabbit tissues and leukocytes thymosin β_4^{Ala} and β_{10} are expressed in equal amounts. Oxidation of the methionyl residue at position 6 of β-thymosins increases the dissociation constant of the complex about 20-fold.[35,37,41] Concomitantly, a 20-fold molar excess of β-thymosin-sulfoxides is necessary to inhibit salt-induced polymerization of G-actin while already equimolar amounts of nonoxidized β-thymosins are sufficient to inhibit polymerization. Comparable to the oxidation of thymosin β_4 truncation of the first 6 or 12 amino acid residues ($\beta_4^{7\text{-}43}$ or $\beta_4^{13\text{-}43}$) increases the dissociation constant by about 20-fold. However, these truncated peptides are no longer able to inhibit polymerization of G-actin. Truncation of the first 23 amino acid residues completely abolishes the interaction with G-actin.[37] Changing the C-terminal structure of thymosin β_4 also modulates the interaction with G-actin. Removal of the last two amino acid residues of thymosin β_{10} results in a minute but measurable decrease in actin binding.[39] After removing the putative α-helix at the C-terminus, the peptide $\beta_4^{1\text{-}30}$ still inhibits formation of F-actin, although a 25-fold molar excess over G-actin was needed. Further deletions yield peptides with even more reduced activities. Peptide $\beta_4^{1\text{-}24}$ inhibits actin polymerization only when present in 50-fold molar excess and peptide $\beta_4^{1\text{-}16}$ missing the putative actin-binding motif (LKKTET) is no longer able to bind to and inhibit G-actin polymerization.[42] These studies clearly indicate that truncating the N-terminus even by only 6 amino acid residues or the C-terminus by more than 2 amino acid residues change the actin sequestering properties. β-thymosins seem to be perfectly designed by nature to sequester G-actin. The molecular mass of β-thymosins (5 kDa) is only about one-tenth of the molecular mass of G-actin (42 kDa) but this relative small protein is able to bind G-actin in a one-to-one ratio with rather low affinity but fast binding kinetics. Even small changes done by evolution in nature or biochemists in their studies may have profound effects on the efficacy of β-thymosins.

While the affinity of β-thymosins for actin is relatively low compared with other monomer-binding proteins, binding kinetics are extremely fast.[35,43] For thymosin β_4 and ATP-G-actin under physiological conditions, an on-rate of 1-2 μM^{-1} sec^{-1} and an off-rate of 3-5 sec^{-1} were reported. While the on-rate was relatively unaffected by ionic strength or by replacement of ATP with ADP, the off-rate was observed to increase as much as two orders of magnitude.[43]

Measurement of the rate and final level of actin polymerization in the presence of thymosin β_4 showed that thymosin β_4 in effect reduces the concentration of monomers available for polymerization, but does not otherwise alter the rate or final extent of polymerization. Thymosin β_4 thus sequesters actin monomers, but unlike other monomer-binding proteins, does not cut, cap, or nucleate actin filaments.[11] In vitro at high concentrations, weak binding of thymosin β_4 to F-actin has been inferred, with the K_d in the millimolar range.[44] The biological significance of this interaction is uncertain, since most F-actin in cells is associated with proteins such as tropomyosin and α-actinin, which are likely to block the binding of thymosin β_4. However clonal cell lines strongly overexpressing thymosin β_{10} had more polymerized actin than control cells and the filaments appeared thicker after staining with fluorescent phalloidin.[45] Thymosin β_4 at 200 μM can be chemically cross-linked to F-actin. In the presence of phalloidin, the chemically cross-linked thymosins β_4 – actin complex can be incorporated into F-actin. Analysis of the helical parameters of F-actin revealed an increase of the crossover spacing of the two right-handed long-pitch helical strands from 36.0 to 40.5 nm, which may imply a change in actin monomer conformation[46] These data on binding of β-thymosins to F-actin indicate that β-thymosins might be not only simple G-actin sequestering proteins and fulfill additional functions in the interplay with other G- and F-actin binding proteins inside the complex environment of a cell.

Thymosin β_4 inhibits the exchange of actin-bound nucleotide by a factor of about 10^{10}.[47] Nucleotide exchange is thought to require at least transient opening of the nucleotide-binding cleft, suggesting that thymosin β_4 stabilizes the closed conformation of the cleft. This conformational stabilization is also reflected in reduced "breathing movements", shown by the inhibition of amide proton exchange.[43]

Structure of the G-Actin-Thymosin β_4 Complex

The three-dimensional structure of the G-actin-thymosin β_4 complex has not yet been determined experimentally, but various structural features have been deduced from spectroscopic and biochemical studies and several models for the structure have been proposed. Circular dichroism and ^{1}H-NMR spectroscopy show that thymosin β_4 alone does not have a unique conformation in aqueous solution, though residues 5-16 tend to form an α-helix.[19,48] Formation of the complex results in increased negative ellipticity at 222 nm, indicating that total α-helix content increases.[16] Isotope-edited NMR shows that this increase corresponds both to the stabilization of the α-helix at residues 5-16 and formation of a smaller α-helical segment at residues 31-39 of thymosin β_4.[23] Actin conformation is also affected by formation of the complex, as shown by changes in the CD spectra of tyrosine and tryptophan chromophores (which are absent in thymosin β_4) and by the increased susceptibility of actin subdomain 2 to proteolysis.[43] A recent study using FRET spectroscopy shows that binding of thymosin β_4 to G-actin also decreases the separation between fluorophores on subdomains 1 and 2.[18]

An initial structural model was proposed on the basis of the early spectroscopic studies and the G-actin-thymosin β_4 contacts identified by cross-linking (Fig. 2A).[49] More recently, two structural models were proposed for the G-actin-thymosin β_4 complex based on the crystallographic structures of G-actin bound to homologs of thymosin β_4 (Fig. 2B,C).[50,51] Hertzog et al reported the structure of G-actin bound to domain 1 of ciboulot, which shares 30% sequence identity with thymosin β_4, while Irobi et al reported the structure of actin bound to a construct consisting of thymosin β_4 residues 21-43 fused to the C-terminus of

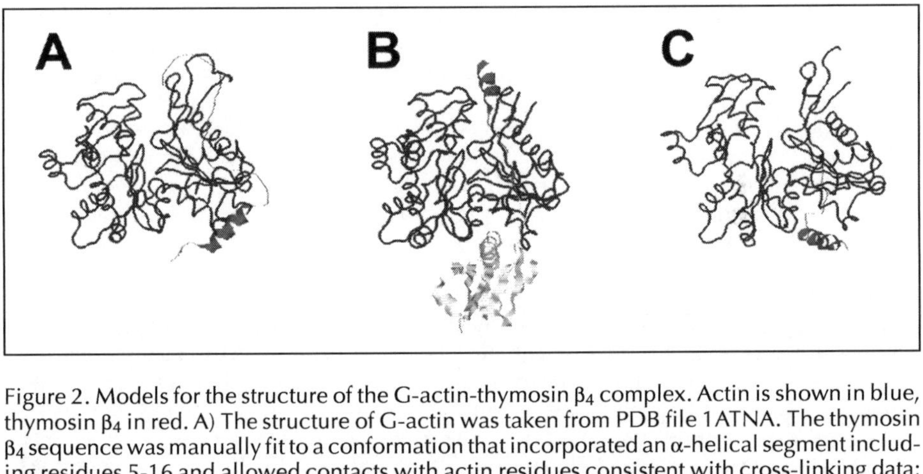

Figure 2. Models for the structure of the G-actin-thymosin β_4 complex. Actin is shown in blue, thymosin β_4 in red. A) The structure of G-actin was taken from PDB file 1ATNA. The thymosin β_4 sequence was manually fit to a conformation that incorporated an α-helical segment including residues 5-16 and allowed contacts with actin residues consistent with cross-linking data; modified from reference 49. B) Crystallographic structure of G-actin bound to the construct consisting of gelsolin residues 27-152 followed by residues 21-43 of thymosin β_4; the gelsolin domain is shown in grey (from PDB file 1T44; Irobi). C) Crystallographic structure of actin bound to ciboulot domain 1 (from PDB file 1SQK; Hertzog). A color version of this figure is available online at www.Eurekah.com.

gelsolin segment 1 (gelsolin residues 27 – 152). Both models for the G-actin-thymosin β_4 complex were based on these crystallographic structures. The N-terminal α-helical region (residues 5-16) of thymosin β_4 contacts the hydrophobic patch on the barbed end of the actin monomer. Residues 21-29 span the nucleotide-binding cleft in an extended conformation and residues 30-40 form an α-helix in contact with the pointed end, though the position of this C-terminal helix is different in the two models.

Both of these homology-based models are generally consistent with the earlier model for the G-actin-thymosin β_4 complex based on cross-linking and spectroscopy: in all three models, the N- and C-termini of thymosin β_4 contact the barbed and pointed ends of the monomeric actin. Cross-linking provided no information about the location of the central region of thymosin β_4. The proposed position in the homology-based models, bridging the two sides of the nucleotide-binding cleft, may provide a structural basis for the inhibition of nucleotide exchange by thymosin β_4. None of these models however should be considered equivalent to a high-resolution structure for the genuine G-actin-thymosin β_4 complex. The sequence differences between ciboulot domain 1 and the N-terminus of thymosin β_4 are sufficient to allow different, though overlapping binding sites. A recent study which compared the structures of actin bound to 3 different WH2 domains (WASP, WAVE, and WIP) showed that despite their high sequence homology, these domains have nonidentical binding sites on actin.[32] In the complex of G-actin with the chimeric construct of gelsolin segment 1 and thymosin β_4 the C-terminus is identical to thymosin β_4, however, the gelsolin domain provides most of the binding energy and may constrain the binding of the thymosin β_4 segment of the construct to favor a binding mode that is less favorable for full-length thymosin β_4.

Extracellular β-Thymosins

Although this review deals only with intracellular β-thymosin, some information about the extracellular β-thymosins will be given. The fact that thymosin β_4 is also found in wound fluid, blood serum and in lower concentrations in blood plasma is quite puzzling. How might the intracellular actin-binding thymosin β_4 be released without a signal peptide? It is conceivable that cells release or lose the 5-kDa thymosin β_4 already at an early stage of damage. When mouse skin was damaged with sodium lauryl sulfate for a short period of time (10 min) for vaccination via skin, it was noticed that thymosin β_4 was released from the skin.[52] Additionally thymosin β_4 is released from blood platelets during aggregation and cross-linked by factor XIIIa to fibrin.[53,54] Recently, it has been found that thymosin β_4 is linked preferentially to the αC domain of fibrin.[55] By virtue of its high intracellular concentration and accessibility of its nine lysine and three glutamine residues, thymosin β_4 is an ideal substrate for transglutaminases. In vitro two out of the three glutamine residues can easily be cross-linked by transgluaminase to fluorescent cadaverines (Oregon Green or dansyl cadaverine). These fluorescent derivatives of thymosin β_4 still bind G-actin and inhibit its polymerization identically to thymosin β_4 and provide a simple way to label β-thymosins.[53] Presently it is not known if β-thymosins are also intracellular substrates of transglutaminases.

Apoptosis and β-Thymosins

During apoptosis of cells, the microfilament system is disrupted. After incubating HL60 cells for 48 hours in the presence of araC the content of thymosin β_4 and actin in the cells decreased to about 30% and 50%, respectively, while the content of total protein remained constant. Already after 24 hours, the amount of mRNA of thymosin β_4 and actin was reduced to 70% and 58%, respectively. Triggering apoptosis has profound effects on transcription of actin and thymosin β_4.[56] Thymosin β_4 overexpressing SW480 colon carcinoma cells have been reported to be more resistant to the cytotoxicity of FasL-bearing T cells and several anticancer drugs.[57,58]

Quantitative Determination and Expression of β-Thymosins

Western blot analyses are often used to determine the amount of β-thymosins. However, thymosin β_4 does not bind very strongly to nitrocellulose. Antibodies against the various β-thymosins are difficult to obtain or available from a few companies at high cost. We have raised specific antibodies against thymosins β_4, β_{10} and β_{15}. These antibodies were generated against C-terminal fragments of the three β-thymosins and are useful for detecting thymosin β_4, β_{10}, or β_{15} chemically cross-linked to actin in Western blot analysis. The epitope recognized by anti-thymosin β_4 are the last four amino acid residues (-AGES). When we used our antibodies or commercially available antibodies against thymosins β_4 for immunocytochemistry, we were not able to abolish the staining by adding large amounts of thymosins β_4 to the primary antibody. Thus, we do not feel confident using the currently available antibodies for immunocytochemistry.

With the advent of the DNA array assays many reports were published showing increases in the expression of the various β-thymosins. In some cases, only probes for one of the various β-thymosins known to be expressed in the cells studied were present on the chip. In none of the published papers, the amounts of β-thymosins were determined on the protein level. Currently only thymosin β_4 can be determined by a commercially available ELISA. However, this ELISA does not discriminate between thymosin β_4 and its sulfoxide. Additionally the ELISA is not very sensitive. The only reliable method to determine β-thymosins is based on extraction of the cells or tissue by 0.4 M perchloric acid followed by reverse phase HPLC and post column derivatization with fluorescamine.[6,41] Using this procedure we can easily detect less than 20 ng of thymosin β_4, identify other β-thymosins present in the cell extract as well as potentially oxidized forms of the peptides at the same time (Fig. 3).

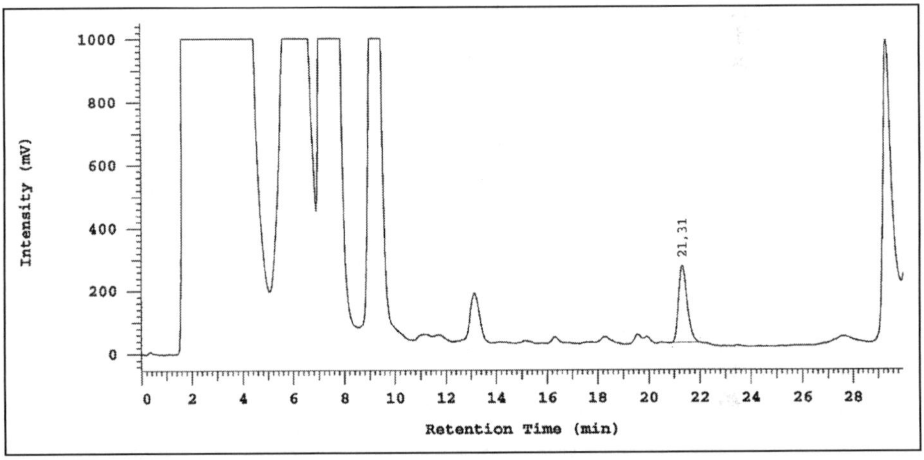

Figure 3. Quantitative determination of thymosin β_4. Human blood (80 µl) was added to 300 µl 0.4 M perchloric acid. After 15 min on ice the precipitated proteins were removed by centrifugation. Thereafter 50 µl of the supernatant solution containing the β-thymosins was added to 300 µl 0.1% TFA. Before injecting the sample 15 µl 1 M NaOH was added to increase the pH of the sample tolerable to the reverse phase column. Separation conditions: Beckman ODS Ultrasphere (250 × 2 mm); linear gradient from 0.1% TFA in 40% acetonitrile within 30 min; detection by post-column derivatization with fluorescamine (for details of the HPLC setup see ref. 62). The peak at 21.31 min corresponds to 100 ng of thymosin β_4 present in about 5 µl human blood. The detection limit of the method is about 10 ng of β-thymosins.

Isolation of β-Thymosins

Various purification schemes have been developed by our laboratories with particular emphasis to avoid proteolysis during purification.[6,16,39,59-62] β-thymosins are soluble in 0.4M perchloric acid or 5 to 10% trichloroacetic acid. In spite of their molecular mass (about 5 kDa), β-thymosins are retained by ultrafiltration using a membrane with 10-kDa cut-off and only recovered in the ultrafiltrate when using a membrane with 30-kDa cut-off. The procedures for purification of β-thymosins consist basically of four steps: (1) extraction of cells or tissue and simultaneous denaturation of proteases, (2) concentrating and desalting by solid-phase extraction, (3) separation of peptides according to their charge (isoelectric focusing, chromatofocusing or ion exchange chromatography) and (4) separation by reverse phase HPLC. The best tissue for isolation of β-thymosins from vertebrates is spleen.

Outstanding Questions

Only a short list of open questions will be given.

- Why do many cells express several β-thymosins? Do they have different functions?
- What is the high-resolution structure of the genuine G-actin-thymosin β_4 complex?
- What is the function of nuclear β-thymosins?
- What other proteins bind to β-thymosins in addition to G-actin?
- What is the function of binding to F-actin at high concentrations of β-thymosins?
- Are β-thymosins substrates of transglutaminases inside of cells?
- How is the expression of β-thymosins regulated?
- What is the half-life of β-thymosins? Is the half-life dependent of the status of the cell?

Acknowledgements

Because of space, not all work on β-thymosins has been included. We have tried to give a short overview and had to omit reports and for this we beg indulgence from the authors.

References

1. Low TL, Hu SK, Goldstein AL. Complete amino acid sequence of bovine thymosin β_4: A thymic hormone that induces terminal deoxynucleotidyl transferase activity in thymocyte populations. Proc Natl Acad Sci USA 1981; 78:1162-1166.
2. Safer D, Elzinga M, Nachmias VT. Thymosin β_4 and Fx, an actin-sequestering peptide, are indistinguishable. J Biol Chem 1991; 266:4029-4032.
3. Goldstein AL, Hannappel E, Kleinman HK. Thymosin β_4: Actin sequestering protein moonlights to repair injured tissues. Trends Mol Med 2005; 11:421-429.
4. Huff T, Müller CS, Otto AM et al. β-Thymosins, small acidic peptides with multiple functions. Int J Biochem Cell Biol 2001; 33:205-220.
5. Wodnar-Filipowicz A, Gubler U, Furuichi Y et al. Cloning and sequence analysis of cDNA for rat spleen thymosin β_4. Proc Natl Acad Sci USA 1984; 81:2295-2297.
6. Hannappel E. One-step procedure for the determination of thymosin β_4 in small tissue samples and its separation from other thymosin β_4-like peptides by high-pressure liquid chromatography. Anal Biochem 1986; 156:390-396.
7. Hannappel E, Xu GJ, Morgan J et al. Thymosin β_4: A ubiquitous peptide in rat and mouse tissues. Proc Natl Acad Sci USA 1982; 79:2172-2175.
8. Xu GJ, Hannappel E, Morgan J et al. Synthesis of thymosin β_4 by peritoneal macrophages and adherent spleen cells. Proc Natl Acad Sci USA 1982; 79:4006-4009.
9. Hannappel E, Leibold W. Biosynthesis rates and content of thymosin β_4 in cell lines. Arch Biochem Biophys 1985; 240:236-241.
10. Hannappel E, van Kampen M. Determination of thymosin β_4 in human blood cells and serum. J Chromatogr 1987; 397:279-285.
11. Weber A, Nachmias VT, Pennise CR et al. Interaction of thymosin β_4 with muscle and platelet actin: Implications for actin sequestration in resting platelets. Biochemistry 1992; 31:6179-6185.
12. Bao L, Loda M, Janmey PA et al. Thymosin β_{15}: A novel regulator of tumor cell motility upregulated in metastatic prostate cancer. Nat Med 1996; 2:1322-1328.

13. Hutchinson LM, Chang EL, Becker CM et al. Use of thymosin β_{15} as a urinary biomarker in human prostate cancer. The Prostate 2005; 64:116-127.
14. Choe J, Sun W, Yoon SY et al. Effect of thymosin β_{15} on the branching of developing neurons. Biochem Biophys Res Commun 2005; 331:43-49.
15. Li X, Zimmerman A, Copeland NG et al. The mouse thymosin β_4 gene: Structure, promoter identification, and chromosome localization. Genomics 1996; 32:388-394.
16. Safer D, Chowrashi PK. β-Thymosins from marine invertebrates: Primary structure and interaction with actin. Cell Motil Cytoskeleton 1997; 38:163-171.
17. Domanski M, Hertzog M, Coutant J et al. Coupling of folding and binding of thymosin β_4 upon interaction with monomeric actin monitored by nuclear magnetic resonance. J Biol Chem 2004; 279:23637-23645.
18. Dedova IV, Nikolaeva OP, Safer D et al. Thymosin β_4 induces a conformational change in actin monomers. Biophys J 2006; 90:985-992.
19. Zarbock J, Oschkinat H, Hannappel E et al. Solution conformation of thymosin β_4: A nuclear magnetic resonance and simulated annealing study. Biochemistry. 1990; 29:7814-7821.
20. Bubb MR, Lewis MS, Korn ED. The interaction of monomeric actin with two binding sites on Acanthamoeba actobindin. J Biol Chem 1991; 266:3820-3826.
21. Vancompernolle K, Vandekerckhove J, Bubb MR et al. The interfaces of actin and Acanthamoeba actobindin. Identification of a new actin-binding motif. J Biol Chem 1991; 266:15427-15431.
22. Van Troys M, Dewitte D, Goethals M et al. The actin binding site of thymosin β_4 mapped by mutational analysis. EMBO J 1996; 15:201-210.
23. Simenel C, Van Troys M, Vandekerckhove J et al. Structural requirements for thymosin β_4 in its contact with actin. An NMR-analysis of thymosin β_4 mutants in solution and correlation with their biological activity. Eur J Biochem 2000; 267:3530-3538.
24. Huff T, Rosorius O, Otto AM et al. Nuclear localization of the G-actin sequestering peptide thymosin β_4. J Cell Sci 2004; 117:5333-5343.
25. Rando OJ, Zhao K, Crabtree GR. Searching for a function for nuclear actin. Trends Cell Biol 2000; 10:92-97.
26. Pederson T, Aebi U. Nuclear actin extends, with no contraction in sight. Mol Biol Cell 2005; 16:5055-5060.
27. Machesky LM, Insall RH, Volkman LE. WASP homology sequences in baculoviruses. Trends Cell Biol 2001; 11:286-287.
28. Paunola E, Mattila PK, Lappalainen P. WH2 domain: A small, versatile adapter for actin monomers. FEBS Lett 2002; 513:92-97.
29. Boquet I, Boujemaa R, Carlier MF et al. Ciboulot regulates actin assembly during Drosophila brain metamorphosis. Cell 2000; 102:797-808.
30. Edwards J. Are β-thymosins WH2 domains? FEBS Lett 2004; 573:231-232.
31. Lappalainen P, Mattila P. Reply to: Are β-thymosins WH2 domains? FEBS Lett 2004; 573:233.
32. Chereau D, Kerff F, Graceffa P et al. Actin-bound structures of Wiskott-Aldrich syndrome protein (WASP)-homology domain 2 and the implications for filament assembly. Proc Natl Acad Sci USA 2005.
33. Jeffery CJ. Moonlighting proteins. Trends Biochem Sci 1999; 24:8-11.
34. Jeffery CJ. Multifunctional proteins: Examples of gene sharing. Ann Med 2003; 35:28-35.
35. Jean C, Rieger K, Blanchoin L et al. Interaction of G-actin with thymosin β_4 and its variants thymosin β_9 and thymosin β_9^{Met}. J Muscle Res Cell Motil 1994; 15:278-286.
36. Carlier MF, Jean C, Rieger KJ et al. Modulation of the interaction between G-actin and thymosin β_4 by the ATP/ADP ratio: Possible implication in the regulation of actin dynamics. Proc Natl Acad Sci USA 1993; 90:5034-5038.
37. Huff T, Zerzawy D, Hannappel E. Interactions of β-thymosins, thymosin β_4-sulfoxide, and N-terminally truncated thymosin β_4 with actin studied by equilibrium centrifugation, chemical cross-linking and viscometry. Eur J Biochem 1995; 230:650-657.
38. Heintz D, Reichert A, Mihelic M et al. Use of bimanyl actin derivative (TMB-actin) for studying complexation of β-thymosins. Inhibition of actin polymerization by thymosin β_9. FEBS Lett 1993; 329:9-12.
39. Huff T, Müller CS, Hannappel E. C-terminal truncation of thymosin β_{10} by an intracellular protease and its influence on the interaction with G-actin studied by ultrafiltration. FEBS Lett 1997; 414:39-44.
40. Yu FX, Lin SC, Morrison-Bogorad M et al. Thymosin β_{10} and thymosin β_4 are both actin monomer sequestering proteins. J Biol Chem 1993; 268:502-509.

41. Huff T, Hannappel E. Oxidation and reduction of thymosin β_4 and its influence on the interaction with G-actin studied by reverse-phase HPLC and postcolumn derivatization with fluorescamine. Anal Chim Acta 1997; 352:249-255.
42. Vancompernolle K, Goethals M, Huet C et al. G- to F-actin modulation by a single amino acid substitution in the actin binding site of actobindin and thymosin β_4. EMBO J 1992; 11:4739-4746.
43. De La Cruz EM, Ostap EM, Brundage RA et al. Thymosin β_4 changes the conformation and dynamics of actin monomers. Biophys J 2000; 78:2516-2527.
44. Carlier MF, Didry D, Erk I et al. Thymosin β_4 is not a simple G-actin sequestering protein and interacts with F-actin at high concentration. J Biol Chem 1996; 271:9231-9239.
45. Sun HQ, Kwiatkowska K, Yin HL. β-Thymosins are not simple actin monomer buffering proteins. Insights from overexpression studies. J Biol Chem 1996; 271:9223-9230.
46. Ballweber E, Hannappel E, Huff T et al. Polymerisation of chemically cross-linked actin: Thymosin β_4 complex to filamentous actin: Alteration in helical parameters and visualisation of thymosin β_4 binding on F-actin. J Mol Biol 2002; 315:613-625.
47. Goldschmidt-Clermont PJ, Furman MI, Wachsstock D et al. The control of actin nucleotide exchange by thymosin β_4 and profilin. A potential regulatory mechanism for actin polymerization in cells. Mol Biol Cell 1992; 3:1015-1024.
48. Czisch M, Schleicher M, Horger S et al. Conformation of thymosin β_4 in water determined by NMR spectroscopy. Eur J Biochem 1993; 218:335-344.
49. Safer D, Sosnick TR, Elzinga M. Thymosin β_4 binds actin in an extended conformation and contacts both the barbed and pointed ends. Biochemistry 1997; 36:5806-5816.
50. Hertzog M, Van Heijenoort C, Didry D et al. The β-thymosin/WH2 domain; structural basis for the switch from inhibition to promotion of actin assembly. Cell 2004; 117:611-623.
51. Irobi E, Aguda AH, Larsson M et al. Structural basis of actin sequestration by thymosin β_4: Implications for WH2 proteins. EMBO J 2004; 23:3599-3608.
52. Huang CM, Wang CC, Kawai M et al. Surfactant sodium lauryl sulfate enhances skin vaccination: Molecular characterization via a novel technique using ultrafiltration capillaries and mass spectrometric proteomics. Mol Cell Proteomics 2005, (MCP published Nov 28, 10.1074/mcp.M500259-MCP500200).
53. Huff T, Ballweber E, Humeny A et al. Thymosin β_4 serves as a glutaminyl substrate of transglutaminase. Labeling with fluorescent dansylcadaverine does not abolish interaction with G-actin. FEBS Lett 1999; 464:14-20.
54. Huff T, Otto AM, Müller CS et al. Thymosin β_4 is released from human blood platelets and attached by factor XIIIa (transglutaminase) to fibrin and collagen. FASEB J 2002; 16:691-696.
55. Makogonenko E, Goldstein AL, Bishop PD et al. Factor XIIIa incorporates thymosin β_4 preferentially into the fibrin(ogen) αC-domains. Biochemistry 2004; 43:10748-10756.
56. Müller CSG, Huff T, Hannappel E. Reduction of thymosin β_4 and actin in HL60 cells during apoptosis is preceded by a decrease of their mRNAs. Mol Cell Biochem 2003; 250:179-188.
57. Wang WS, Chen PM, Hsiao HL et al. Overexpression of the thymosin β_4 gene is associated with increased invasion of SW480 colon carcinoma cells and the distant metastasis of human colorectal carcinoma. Oncogene 2004; 23:6666-6671.
58. Hsiao HL, Wang WS, Chen PM et al. Overexpression of thymosin β_4 renders SW480 colon carcinoma cells more resistance to apoptosis triggered by FasL and two topoisomerase II inhibitors via down-regulating Fas and up-regulating Survivin expression, respectively. Carcinogenesis 2006; 27(5):936-984.
59. Hannappel E, Davoust S, Horecker BL. Isolation of peptides from calf thymus. Biochem Biophys Res Commun 1982; 104:266-271.
60. Safer D, Golla R, Nachmias VT. Isolation of a 5-kDa actin-sequestering peptide from human blood platelets. Proc Natl Acad Sci USA 1990; 87:2536-2540.
61. Cassimeris L, Safer D, Nachmias VT et al. Thymosin β_4 sequesters the majority of G-actin in resting human polymorphonuclear leukocytes. J Cell Biol 1992; 119:1261-1270.
62. Huff T, Müller CSG, Hannappel E. HPLC and postcolumn derivatization with fluorescamine. Isolation of actin-sequestering β-thymosins by reversed-phase HPLC. Anal Chim Acta 1997; 352:239-248.
63. Mihelic M, Voelter W. Distribution and biological activities of β-thymosins. Amino Acids 1994; 6:1-13.
64. Stoeva S, Horger S, Voelter W. A novel β-thymosin from the sea urchin: Extending the phylogenetic distribution of β-thymosins from mammals to echinoderms. J Pept Sci 1997; 3:282-290.
65. Hannappel E, Huff T. The thymosins: Prothymosin α, parathymosin, and β-thymosins. Structure and function. Vitam Horm 2003; 66:257-296.

Chapter 7

Multirepeat β-Thymosins

Marleen Van Troys,* Stien Dhaese, Joël Vandekerckhove and Christophe Ampe

Abstract

Multirepeat β-thymosins contain multiple copies of the β-thymosin actin binding module. This family is mainly distributed within lower metazoan species and, with one exception, absent in mammals in which the classical single repeat β-thymosins appear dominant. The repeated nature in combination with sequence variation in the consecutive modules renders these proteins different actin modulating capacities as compared to the classical β-thymosins. These properties are discussed in function of recent structural models indicating how these proteins contact actin. The importance of the multirepeat β-thymosins is underscored by their crucial role in neuronal development and reproduction.

Introduction

Actin filament turnover forms the basis of essential cellular properties ranging from cell division to cell migration.[1,2] Consequently actin dynamics are crucial during embryogenesis and morphogenesis of eukaryotic organisms. The dynamic equilibrium between actin monomers (G(lobular)-actin) and polymeric filamentous (F)-actin can be viewed as a cyclic process in which actin monomers, loaded with ATP and a divalent cation, associate with the fast growing (barbed or plus) end of an actin filament. Subsequently associated monomers undergo ATP-hydrolysis and finally dissociate at the other (pointed or minus) filament end as ADP-actin. At equilibrium, the kinetic and structural differences present at both filament ends drive this cyclic process in a unidirectional fashion also called treadmilling.[3] It is evident that changes in the number of free barbed filament ends or changing concentrations of polymerization competent actin monomers will shift the G/F-equilibrium.[4,5] In cells, this system is strongly regulated by the activity of actin binding proteins with diverse functions such as monomer sequestration, barbed end elongation or barbed end capping.[1,6]

The multirepeat β-thymosins were first reported in 1999[7] and more members are detected as genome sequencing projects are being pursued. As their name suggests they share homology with β-thymosins and thus are actin binding proteins. We here present their distribution within the eukaryotic kingdoms and their structural and biochemical properties in relationship to other actin binding proteins such as the single repeat β-thymosins and WH2-domain containing proteins.[8] We discuss in more detail ciboulot from fruitfly,[9] the amoebal actobindin,[10] tetraThymosinβ from nematode[11] and CSP(condition stimulated phosphoprotein)-24 from sea slug[12] since these have been extensively studied both biochemically and/or in vivo.

*Corresponding Author: Marleen Van Troys—Department of Biochemistry, Ghent University, A. Baertsoenkaai 3, B-9000 Gent, Belgium. Email: leen.vantroys@ugent.be

Actin-Monomer-Binding Proteins, edited by Pekka Lappalainen. ©2007 Landes Bioscience and Springer Science+Business Media.

The Multirepeat β-Thymosin Family: Evolutionary Relationships and Structural Characteristics

The multirepeat β-thymosin proteins contain multiple copies of the β-thymosin actin binding module.[7] This characteristic, together with functional features (see below), sets them apart from the classical β-thymosins that consist of just one of these modules. The number of copies or repeats varies from two to no less than twenty-eight in the recently discovered and highly intriguing thypedin protein from the primitive metazoan *Hydra vulgaris*.[13]

Figure 1A illustrates the taxonomic coverage of this family compared to the one of the classical β-thymosins. Two double repeat members in amoeba and one in *Dictyostelium discoideum* represent the multirepeat β-thymosins in Protista. The other proteins are found in Metazoa in six out of the eight phyla (Fig. 1A, left). Multirepeat β-thymosins appear absent from Fungi and plants, although nonannotated EST-sequences with significant similarity to actobindin are found in *Chlamydomonas*, soybean and sorghum. The β-thymosin distributions (Fig. 1A) are in strong contrast to that of the in sequence partly similar WH2-repeat proteins that are ubiquitously expressed in all eukaryotic kingdoms. Based on current datasets the multirepeat proteins are the only beta-thymosin homologs in Cnidaria (thypedin,[13] 28 repeat-protein), in flat and round worms (two and four repeat proteins respectively, the latter termed tetraThymosinβ)[11] and in Arthropoda (typified by ciboulot,[9] three repeats) (Fig. 1A, left). The single repeat β-thymosins are not represented in these phyla (Fig. 1A, right). Different species of Mollusca however possess either a single or double repeated variant.[14] The repeat β-thymosins are largely absent from Echinodermata and Chordata (Fig. 1A, left) whereas classical β-thymosins are broadly expressed here (Fig. 1A, right). We only found a five repeat protein in the sea squirt *Ciona intestinalis,* which belongs to the Urochordata, the most atypical of the Chordata subphyla. In addition, a neuroblastoma like β-thymosin and its nearly perfect duplication are together present in mice. We isolated the transcript of this double repeat in several types of adult mouse tissue (Dhaese et al, unpublished). As it is not present in related species, even not in other rodents, it probably arose recently in evolution. The latter example also illustrates that the occurrence of multirepeat and single repeat proteins is not necessarily mutually exclusive.

For some multirepeat β-thymosins, splice variants are present (see alignment discussed below). Alternative splicing of the transcript coding for the three repeat protein ciboulot from *Drosophila melanogaster*, results in a two-repeat variant with a longer second repeat that differs carboxyterminally. Based upon exon skipping, a four repeat and five repeat protein, respectively termed CSP-24 and CSP-29, are present in *Hermissenda crassicornis*.[12] In *Apis mellifera* two different splice variants with three and four repeats are found.

The evolutionary path that resulted in this distribution of the different β-thymosins is still elusive: the multirepeat β-thymosins appear more frequently in lower eukaryotes and in protista; the single forms are clearly more typical for Echinodermata and vertebrates (Fig. 1A). On the other hand, single repeats are already present in primitive eukaryotes such as Porifera (sponges).[15] Taken together, this suggests that after ancient gene multiplication events the multirepeat proteins evolved and that the single repeat β-thymosins in mammals were subsequently generated by partial deletion from these longer variants.

Both from an evolutionary and from a functional viewpoint, it is important to clearly define the features of the β-thymosin module (InterPro-domain 001152 or Pfam-domain PF01290) to allow distinguishing it from the shorter WH2-domain (InterPro 003124, Pfam domain PF02205). The latter are present in a very large group of eukaryotic proteins and in a few cases also as repeated domains.[8,16,17] We illustrate the features of the β-thymosin module in Fig.ure1B using human thymosinβ4.[18,19] The central part of the module is a strongly conserved tetrapeptide sequence (17-LKKT-20 in thymosinβ4)[18] starting and ending with a hydrophobic or noncharged residue and containing one or two positively charged residues.[19,20] This consensus motif is preceded by a stretch of amino acids forming an amphipathic α-helix

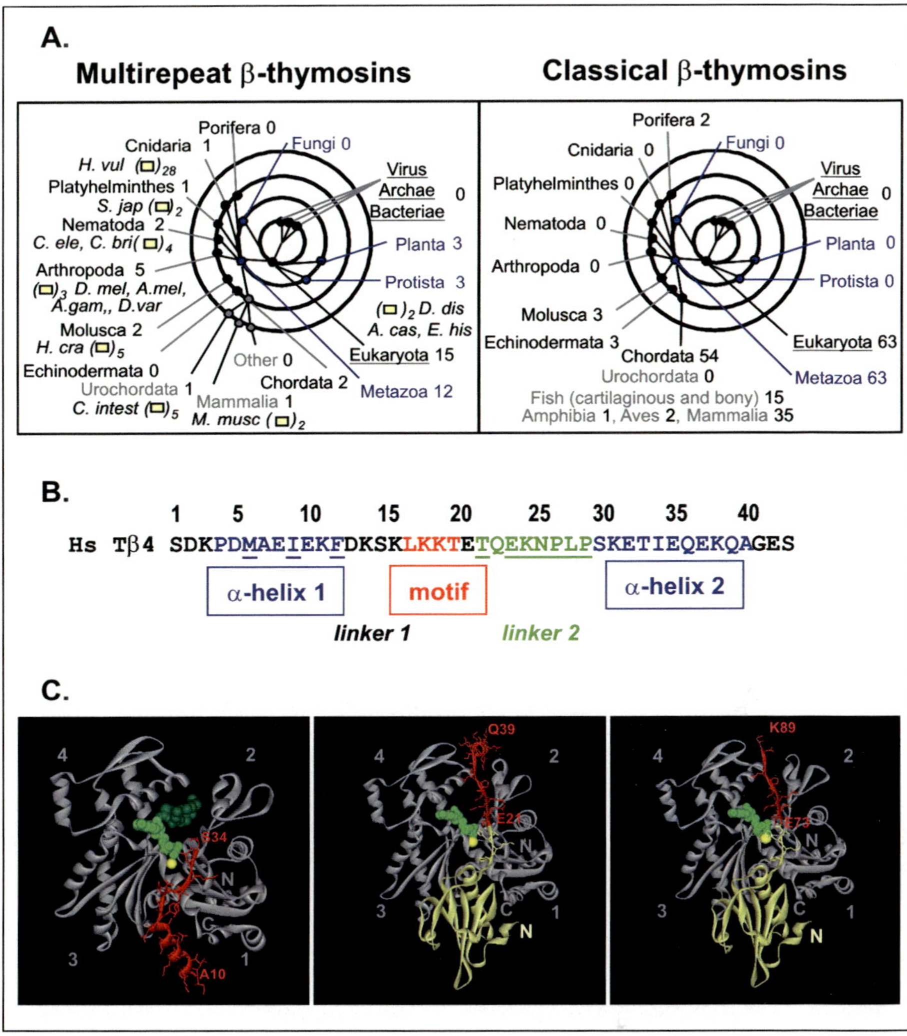

Figure 1. Taxonomic distribution and structural properties of multirepeat β-thymosins. A) Taxonomic coverage of the multirepeat β-thymosin proteins (numbers in bold) with indication of the number of repeats (numbers in subscripts after yellow blocks) (left) and of the classical, one repeat, β-thymosins on the right; representation adapted from InterPro site. For species names see legend to Figure 2. B) Structural features of the β-thymosin actin binding module as present in human thymosinβ4 (NP_066932) (see text for details). C) X-ray based structure of the complexes of monomeric actin with ciboulot[23] (left, 1SQK, residues 10-34 of ciboulot repeat 1 are visible), with a hybrid consisting of gelsolin residues 27-152 and the C-terminal part of thymosinβ4 (residues 21-43)[28] (middle, 1T44, residues 21 to 39 of thymosinβ4 are visible and indicated) and with an analogous hybrid in which the gelsolin sequence is fused to ciboulot residues 72-129 (right, 2FF6, ciboulot residues 73-89 are visible and indicated).[29] The gelsolin part of the hybrid maintains actin in the monomeric state.[40] N, C respectively indicate the amino- and carboxytermini of the actin molecule or gelsolin hybrid, subdomains 1 to 4 of actin are indicated. Actin is in grey, the gelsolin part of the hybrid in light yellow, ciboulot repeat 1 (left) or 2 (right) repeat residues and β-thymosin residues (middle) are in red; actin associated divalent kation, ATP/ADP and latrunculin are in dark yellow, light and dark green respectively.

and containing a hydrophobic triplet (hxxhxxh). C-terminally of the central motif is a stretch of amino acids that can be defined by a (loose) consensus sequence: 'x[T/S/V/P]x[E/D/V/I]KN[P/T]L' (linker 2 in Fig. 1B) connecting to a second α-helix initiating at a proline (residues Pro-29 until Ala-40 in thymosinβ4). We note that in contrast to β-thymosin proteins (single and multirepeat), the sequence C-terminally of the motif is not conserved in WH2-repeats.

Using amino acid substitutions we have been able to pinpoint residues in thymosinβ4 that are essential in actin interaction. We showed the contribution to actin binding of three hydrophobic amino acids (Fig. 1B) and of positively charged residues in the aminoterminal helix 1 as well as the crucial role of residues 1 and 2 of the motif.[18,19,21] In addition, little mutational tolerance is allowed at position four of the motif.[20] These data derived from mutagenesis studies are corroborated by 3D-structural studies (NMR and X-ray diffraction) on single or multirepeat β-thymosins in free or actin monomer bound conditions. Like β-thymosins,[22] the larger proteins do not adopt a single stable fold in the uncomplexed state (ciboulot, tetraThymosinβ)[11,23] and, as in thymosinβ4,[24] actin binding is proposed to stabilize a unique conformation. Crystals formed by the three-repeat ciboulot from *D. melanogaster* and one actin molecule, allowed visualizing the interface between the N-terminal α-helix of repeat 1 of ciboulot and the hydrophobic cleft formed by actin subdomain 1 and 3 as well as the interaction of the central motif of repeat 1 near the nucleotide binding loops in actin (Fig. 1C, left panel).[23] The central motif contacts actin in an extended conformation, as already predicted by binding studies on thymosinβ4-mutants.[21,25] The corresponding part of the WH2-domain was recently shown to follow the same path.[26] The suggested contact site of the second half of the β-thymosin repeat in actin lies across subdomain 2 and extends towards the top of this subdomain where it may putatively interfere with an actin-actin contact in F-actin. This was shown for the classical thymosinβ4 by NMR[27] and by X-ray crystallography[28] (Fig. 1C, middle panel). Recently the interface of the C-terminal half of ciboulot repeat 2 was determined: it follows the same path as the second half of thymosinβ4 except that the C-terminal sequence (corresponding to helix 2 in thymosinβ4) is not ordered (Fig. 1C, right).[29]

How well the binding models, arising from combining the cib1-structures (N-terminal half including the tetrapeptide motif) either with that of thymosinβ4 or cib2 (C-terminal half) in Figure 1C, will be applicable to each of the different modules in multirepeat proteins will depend on their degree of similarity to the typical β-thymosin sequence. The primary structures of all family members are aligned in Figure 2. We introduced repeat-boundaries, in part based on the strongly conserved nature of the 28-repeat thypedin protein displaying higher identity between alternate repeats (i.e., 1 with 3, 5 etc and 2 with 4, 6 etc.) and in part on alignment with the thymosinβ4 sequence. With the exception of the thypedin protein and the *Mus musculus* double repeat, the repeats within one protein have significantly diverged in primary structure: interrepeat sequence identity varies between 37 and 51% in the related *Hermissenda* five-repeat protein CSP29, between 25 and 46% in *Caenorhabditis elegans* tetraThymosinβ[12] and between 47 and 67% in *D. melanogaster* ciboulot.[9] As reported by Crow et al,[12] this degree in similarity is largely retained upon comparing repeats derived from these different proteins. The alignment in Figure 2 demonstrates the conservation in this family of the tetrapeptide motif, of the second linker sequence, of the start of the second α-helix and of a number of charged residues within this helix. The two repeat variants, especially those from protists deviate most strongly in primary structure from the consensus homology sequence. The amino acids constituting the hydrophobic patch in the N-terminal α-helix are also largely conserved although more mutational tolerance is observed here. Putatively some plasticity is possible in fitting this helix in the hydrophobic groove on actin. The observed sequence differences may of course affect the affinities and even function of the different repeats.

```
Hs thymβ4(1-43)                 SDKPDMAE-IEKFDKSKLKKTETQEKNPLPSKETIEQEKQAGES*

Hv thypedin(1-43)              MEEAVATRAEVKTFDQSNLKHVETEEKNPLPTAATLREELRPE
C. int(85-127)                 MADKPKMTE-VTEFDKNKLKGTQTQEKQALPTKEQLMDEKSRA
Hc Csp29(1-42)               MSAAPQGLMSDLT--KEAKLKSVETVEKNPLPTAEAIKDEKQHQ
Hc Csp24(1-33)               MSAAPQGLMSDLT--KEAKLKSVETVEKNPLPTAE----------
Ce tTβ(1-48)           MAAVTELPKMNQELAGAVR--EGLELKKVETTEKNVLPTKEDVAEEKQHV
C. bri(1-48)           MAAVTELPKMNQELAGAVR--EGLELKKVETNEKNILPTKEDVEVEKQHV
A. mel A(1-52)        MSSPSLKDLPKVALDLKSELEGFNHGCMKKASTAEKNVLPSAEDVRQERQHS
A. mel B(1-44)        MSSPSLKDLPKVALDLKSELEGFNHGCMKKASTAEKNVLPSAED--------
Dm cibA(1-54)       MAAPAPALKDLPKVAENLKSQLEGFNQDKLKNASTQEKIILPTAEDVASEKAQR
Dm cibC(1-54)       MAAPAPALKDLPKVAENLKSQLEGFNQDKLKNASTQEKIILPTAEDVASEKAQR
A. gam(1-53)         AGQESTPASYPRVKPDFKSELESFRTETLAKADTQEKNCLPTAADVQAEKTQQ
D. var(1-47)               MDKHPKVADEIQQELASFNAASLKHTETQEKVLLPSKEDVQQEKIHN
S. jap(1-42)                   MADAIKVLEDIDGFDKQKLRHVETEEKVVLPDKEVIAKEKTE
M. musc(1-41)                  MSDKPDLSE-VETFDKSKLKKTNTEVKNTLPSNENKMSDKPD

Hv thypedin(44-79)                VLPDVSEVEKFDASKLKSVTPEVKCHLPTKDVIQEE...
C. int(128-155)                         EAASYDHSQMKHVEPAEKNTLKDDYLRE
Hc Csp29(43-80)                     DHIDTISNFRRASLKKSESVEKSNLPSLAAISQERSQ
Hc Csp24(43-50)                     -----------------------------AISQERSQ
Ce tTβ(49-86)                       ERIHEIEHFDSTKLHSTPVKEKIVLPSADDIKQEKQHL
C. bri(49-86)                       ERIHEIESFDSTKLHSTPVKEKVVLPSAEDIKQEKQHQ
A. mel A(53-89)                     ELIHDVETFKPDQLKHADTKEKIILPNAKDVAAEKTQQ
A. mel B(45-52)                     ------------------------------VAAEKTQQ
Dm cib A(55-91)                     SIFEGITAFNQNNLKHTETNEKNPLPDKEAIEQEKEKN
Dm cib C(55-97)                     SIFEGITAFNQNNLKHTETNEKNPLPDKEGEGEESVHRRHREL*
A. gam(54-90)                       SVIEGIEGFDASRLKHAETKEKNPLPDVEAIQAEKGVQ
D. var(48-84)                       SLLEGVEQFEKTSMKHAQTQEKVCLPKKEDIESEKEHK
S. jap(43-77)                         KQLLQEIETPPSKHTSTKEKN-LPTKDDIVAEKAIH*
M. musc(43-81)                       LSE-VETFDKAKLKKTNTEVKNTLPSKETIQQEKEHNERT*

C. int(156-172)                     KLPGQVASVNKESLKHVETQEKNPLPSGADIAIERAP
Hc Csp24(81-117)                    DVRERIGSFNKDELKKTDTSEKTVLPSIDDIGQEKKE
Hc Csp24(51-87)                     DVRERIGSFNKDELKKTDTSEKTVLPSIDDIGQEKKE
Ce tTβ(87-143)                      ELTDKINNFPSENLKKTETIEKNVLPSPTDVAREKTL
C. bri(87-143)                      ELTDGIQNFPSENLKKTETTEKNVLPSPTDIAREKTL
A. mel A(90-123)                    TLMSGIETFDPSSLKHTETQEKNLLPDIQQ---EKGK
A. mel B(53-89)                     TLMSGIETFDPSSLKHTETQEKNLLPDMDAIQQEKGK
Dm cib A(92-129)                    QFIAGIENFDAKKLKHTETNEKNVLPTKEVIEAEKQA*
A. gam(91-127)                      QFIAGIESFDTKSLKHADTVEKNLLPTAETIEAEKRA*
D. var(85-122)                      QMIEGIETFDPSKLKHAETSVKNPLPTKEVIEQEKAA*

C. int(173-205)                         SDAANFPHEKLKHVETKEAQVLPSKEDVAQEKT
Hc Csp29(118-153)                  VALKESISGFDKSNLKHSEVVEKNSLPPQEAVETEKK
Hc Csp24(88-123)                   VALKESISGFDKSNLKHSEVVEKNSLPPQEAVETEKK
Ce tTβ(144-171)                         QMAASFDKSALHHVETIVSTDVRVTEAQ*
C. bri(144-171)                         QMAASFDKSALHHVETVVSNDVRVTDAQ*
A. mel A(124-155)                  QQLISGIENFDPAKLKHAETLEKNPLPTKEGK*
A. mel B(90-127)                   QQLISGIENFDPAKLKHAETLEKNPLPTKEAIDAEKIA*

C. int(206-237)                        IRDIGLFKRDSLNKIETQEKNTLPTKETIEEE*
Hc Csp29(154-189)                 ENEFRKSIEAFPKEGLKKTECAEKNTLPTKETIQAE*
Hc Csp294(124-159)                ENEFRKSIEAFPKEGLKKTECAEKNTLPTKETIQAE*


A. cas(1-35)                      MNPELQSAIGQG-AALKHAETVDKSA-PQIENVTVKK
D. dis(1-38)                   MSTTANPLLAEINKG-TDLKHAETQDKSA-PIIENVPIKK
E. his(1-33)                      DAKALAGIAD---AKLKHTETGDKSA-PVIENVEIKK

A. cas(36-88)                     VDRSSFLEEVAKPHELKHAETVDKSG-PAIPEDVHVKKVDRGAFLSEIEKAAKQ*
D. dis(39-90)                     NDHSSLLGEVEKGAQLKHVETQDRSA-P-VTEGATVKSNNHSALLGEIKSKAQE*
E. his(34-85)                     GDRNELLSGIKEGKELKKAETNDRSA-PVIPADAKVQEDNRGALLADIQATAK*
```

Figure 2. Sequence alignment of multirepeat β-thymosin proteins. The alignment emphasizes the repeated structure of the proteins. Proteins displayed are NP_066932 (thymosinβ4) from *Homo sapiens* (Hs), AAW82079 (thypedin) from *Hydra vulgaris* (Hv) (only repeats 1 and 2 are shown), BAC57524 from *Ciona intestinalis* (minus first 85 residues), splice variants AAN08023 (Csp-29) and AAN08025 (Csp-24) from *Hermissenda crassicornis* (Hc), NP_509430 (tetraThymosinβ, tTβ) from *Caenorhabditis elegans* (Ce), BP:CBP05424 from *Caenorhabditis briggsae*, splice variants XP_623926 and a non-annotated transcript from *Apis mellifera*, splice variants AAN09116 (ciboulot A) and AAF45919 (ciboulot C) from *Drosophila melanogaster* (Dm), XP_310894 from *Anopheles gambiae* str. PEST, AAO92284 from *Dermacentor variabilis*, AAX30141 from *Schistosoma japonicum*, NP_997150 from *Mus musculus*, P18281 or actobindin from *Acanthamoeba castellanii*, XP_646040 from *Dictyostelium discoideum*, O15602 from *Entamoeba histolytica*. Numerous homologs in invertebrate animals where (partial) EST information is available are not taken up. Strongly similar residues within the conserved features (see Fig. 1B) are in blue, the hydrophobic patch (as found in thymosinβ4)[19] is boxed; numbers indicate the sequence segments of the particular proteins, * indicates the carboxyterminus of the proteins. Because of stronger divergence the Protista homologs are shown separately at the bottom.

Biochemical Studies Reveal Functional Versatility within the β-Thymosin Repeats in Affecting Actin Dynamics

The classical β-thymosins are actin monomer sequestering proteins that are able to shift the dynamic equilibrium between G- and F-actin towards the monomeric form (Fig. 3A). In cells they are the major sequestering proteins and believed to control the monomer reserve in a concentration dependent fashion.[30] As expected from their high similarity with these classical β-thymosins, full-length multirepeat β-thymosin proteins as well as their individual repeats interact with actin monomers. This has been shown via various experimental approaches for *D. melanogaster* ciboulot,[9,31] for *Acanthamoeba castellanii* actobindin[32,33] and for *C. elegans* tetraThymosinβ.[11] For the Csp24 repeat protein from *H. crassicornis* and for thypedin from *H. vulgaris* actin binding was confirmed using coimmunoprecipitation from total tissue lysates.[12,13] Monomer binding results in inhibition of nucleotide exchange on the actin monomer (demonstrated for ciboulot and actobindin) similar as was reported for single repeat β-thymosins.[31,34] Ciboulot and actobindin also display preferential binding to ATP-bound actin monomers in line with the binding or sequestration of polymerization competent monomers.[31]

As stated above, also individual repeats of the multirepeat β-thymosins interact with actin monomers. We demonstrated this for all four repeats of tetraThymosinβ that were studied under form of chemical peptides and within full-length variants in which three out of four repeats carried a detrimental motif-mutation.[11] Also repeat 1 as well as repeat 2 and 3 of ciboulot interact with G-actin.[31,26] TetraThymosinβ interacts, under G-actin binding conditions, simultaneously with at least three actin monomers.[11] Also for actobindin a stoichiometry of one with two actin monomers has been reported.[32,33] Binding affinities obtained for ciboulot and tetraThymosinβ do not differ substantially from affinities derived for single repeat proteins suggesting the absence of cooperativity in monomer binding. Based on these biochemical data and recent structural data,[29] one can extrapolate that the multirepeated nature provides all members of this family the capacity to interact simultaneously with multiple monomers. The elongated three-dimensional structure of these proteins is able to accommodate for this, as also suggested by in silico modeling.[23,26,28,29]

Intriguingly, we have been able to show that 2 out of four repeats of *C. elegans* protein tetraThymosinβ and the full-length protein interact with the polymerized form of actin.[11] Repeats that display F-actin binding are no longer able to sequester actin monomers (in the presence of capped barbed filament ends (see also below)). We demonstrated this using isolated repeats and full-length variants in which three out of four repeats where inactivated by mutations within the motif (see Fig. 8 in ref. 11). The F-actin binding activity is most evident for repeat 3 that, surprisingly, is highly similar in sequence to thymosinβ4. Analogously, Aguda et al[29] recently reported that also ciboulot repeats 2 and 3 display the capacity to interact with actin filaments.

In studying the actin monomer sequestering capacity of these proteins, a major difference with the classical β-thymosins was observed. The latter prevent polymerization of bound actin monomers independent of the status of the barbed ends of the actin filaments they are incubated with. However Boquet et al[9] reported that ciboulot only efficiently sequesters actin when barbed end of actin filaments are capped. This suggests that actins, bound to ciboulot, are competent to associate with a free barbed end (Fig. 3A). This activity is reminiscent of the activity of another actin monomer binding protein profilin.[35] Profilin-actin complexes are considered as actin monomer-delivery systems for free barbed ends. Upon binding the barbed end, profilin dissociates. An limited sequestering capacity and elongation-promoting function has also been demonstrated for actobindin[31] and for tetraThymosinβ[11] using the same sequestering assay.

In light of the biochemical experiments it is logical to correlate the functional dissimilarity between these multirepeat proteins and the single repeat β-thymosins to their multirepeated nature. For tetraThymosinβ, we proposed a model that underlies the elongation promoting activity as a result of cooperative action of all four repeats (Fig. 3B).[11] This model is based on the stoichiometry of the complex with actin, the observed preferential F-actin-binding of certain repeats (especially repeat 3) and on the fact that full elongation promoting activity of

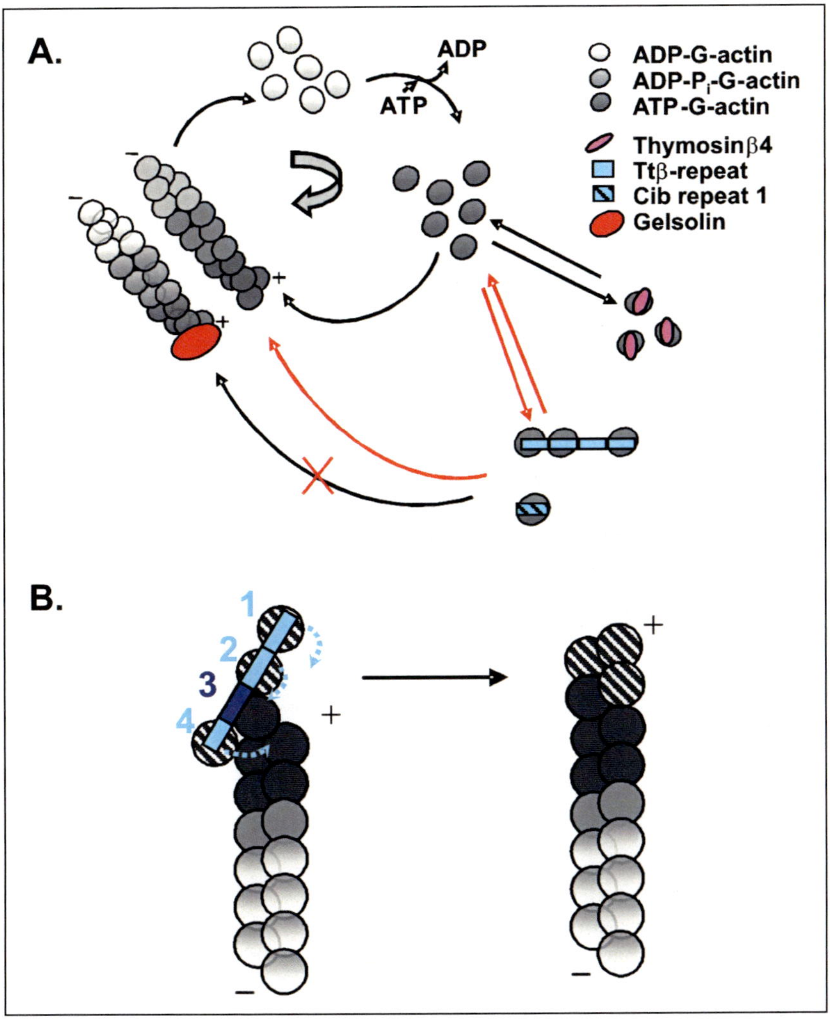

Figure 3. Effects of multirepeat β-thymosins on actin dynamics. A) Actin polymerization cycle showing the dynamic equilibrium between G- and F-actin and the associated nucleotide hydrolysis and exchange steps. Thymosinβ4 sequesters ATP-actin monomers. By contrast multirepeat β-thymosins bind multiple actin monomers and also have the capacity to deliver the bound actin monomer to a free barbed end, thereby promoting barbed end elongation, but not to a capped barbed filament end. TetraThymosinβ and ciboulot repeat 1 are shown as representative for the activities of the multirepeat proteins (see text for details). B) Proposed mechanism by which tetraThymosinβ assists in filament elongation. TetraThymosinβ repeat 3 (darker blue) is contacting the filament, whereupon actin monomers bound to repeats 1, 2 and 4 can be transferred to the filament. Color code is as in (A).

tetraThymosinβ is critically dependent on repeat 3 and only displayed by the full-length protein having four active repeats.[11] In brief, tetraThymosinβ binds actin filaments via repeat 3 and in this way brings the actin monomers associated to the other repeats (1,2,4) in close proximity to the growing barbed end (Fig. 3B). A mechanism of this kind is expected to strongly contribute to the efficiency of barbed end elongation. How release of actin bound molecules is mediated is presently unknown, although G/F-transitions in the bound actin may play a role. In addition, the binding of the aminoterminal helix of one repeat may modulate the binding by the carboxyterminal helix of the previous repeat.[28,29] This same mechanistic model, with some adaptation, is also proposed for actobindin and ciboulot by Aguda et al[29] including a concept, based on in silico modeling, that multiple repeats may associate with longitudinally related actin protomers within a filament. We note that this activity model bears similarity to the one recently suggested for the nucleation activity of spire, a protein that contains four WH2-repeats in tandem.[36]

Interestingly, Hertzog et al[23,31] have established that the first repeat of ciboulot has inherent capacity to promote actin filament elongation, even isolated from the rest of the ciboulot protein. They were able to show that specific residues in the motif and in the secondary α-helix of ciboulot repeat 1 are important in this activity since substituting these amino acids on the corresponding positions in thymosinβ4 attributes the barbed end elongation promoting activity to thymosinβ4.

Based upon recent structural data for the β-thymosin module and the related WH2-module[23,26-28] it has been suggested that monomer sequestering activity versus the capacity to function in nucleation or in elongation promoting activity by a β-thymosin module or a WH2-domain is mainly related to the binding affinity of the aminoterminal helix, to the orientation of the β-thymosin linker 2 on actin subdomain 2 and to the binding of the C-terminal β-thymosin α-helix at an actin-actin-interface.[26] This appears valid when comparing classical β-thymosins with typical WH2-domain proteins that lack strong conservation carboxyterminally of the motif. However when considering the activity of the modules of the multirepeat β-thymosins in relation to their sequence identity with the single β-thymosins, the function predictive potential based on primary structure becomes less evident. Atypical repeats (e.g repeat 4 in tetraThymosinβ) still sequester monomers whereas a highly conserved repeat (e.g., repeat 3 in tetraThymosinβ) no longer sequesters.[11] The minimal substitutions allowing the sequestration/profilin-like activity switch for a single repeat[23] also demonstrate that subtle changes in the motif of and/or following the motif can drastically modulate the function of this actin binding domain. Most likely the functionality is highly dependent on the motif sequence (our unpublished data) and to the extent in which the motif- and linker 2-contact near the actin hinge region affects the conformation of the bound monomer, as also suggested by recent studies by Dedova et al.[37] Based hereupon, monomer sequestering could even occur without a β-thymosin-actin contact directly competing with the actin-actin contact at subdomain 2.

The current data on multirepeat β-thymosins already reveal that a high level of versatility is possible within β-thymosin module and indicate that we are still far from being able to predict function from sequence, even for this fairly homologous family of proteins. Moreover, the interplay or synergistic action of the multiple repeats, as we showed for tetraThymosinβ, will be important in understanding the functionality of these proteins.

Multirepeat β-Thymosins Are Essential in Neuronal Development and Plasticity and/or in Reproduction

For a number of multirepeat β-thymosins localization and effects of absence of the protein have been determined in vivo. TetraThymosinβ is expressed throughout worm life and found enriched in regions of high actin dynamics.[11] In embryos, it is recruited to the developing primary nerve bundle (termed nerve ring) concomitant with the appearance of actin rich structures at this site (Fig. 4A). Interestingly, tetraThymosinβ is absent from this neural structure in adults indicating it is only present at times that actin polymerization is effectively required. A striking feature is the colocalization of tetraThymosinβ in the cortex of oocytes present in distal ends of adult gonads suggesting a role in oocyte maturation (Fig. 4B). This cortical actin in the oocyte is hypothesized to be important for resisting mechanical stress during the ovulation process. TetraThymosinβ$^{-/-}$ animals indeed display strong defects in oocyte maturation as shown in Figure 4C (right panel) are misshapen (dumpy phenotype) and die as young adults. They probably survive to adulthood due to the presence of protein of maternal origin in the embryo.[11]

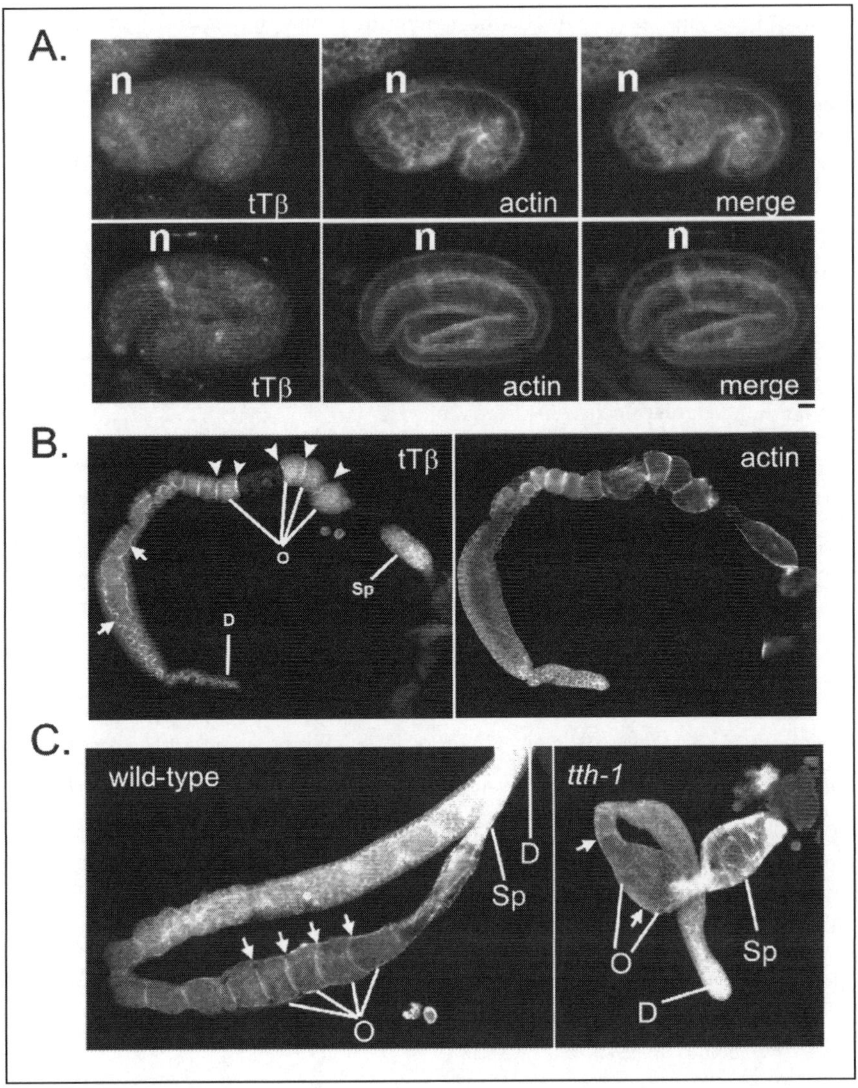

Figure 4. In vivo role of *C. elegans* tetraThymosinβ. A) Immunolocalization of tetraThymosinβ (left) and actin (middle) at different phases of embryonic development (scalebar, 10 mm). In the right panel, colocalization is visualised (red: tetraThymosinβ, green: actin, blue: DAPI-staining). 'n' indicates the site of the nerve ring. B) Immunolocalization of tetraThymosinβ (left panel) and actin (right) in the dissected adult gonad (D, distal tip; Sp, spermatheca; O, oocytes) (scalebar, 50 mm). TetraThymosinβ localizes to the cytoplasm and the cortex (arrowheads in tetraThymosinβ staining) of the oocytes and to the inside edges of the membrane cubicles (arrows in tetraThymosinβ staining) surrounding germ cell nuclei. C) Dissected adult gonads (D, distal tip; Sp, spermatheca; O, oocytes) from a WT (left panel) or *tth-1* (tetraThymosinβ encoding gene) homozygote mutant (right panel) were stained for actin (scalebar, 50 mm). Actin is accumulated at the cortex of the oocytes (left, arrows) in WT but absent in the *tth-1* mutant (right, arrows). In the *tth-1* mutant, deformed oocytes are found in the spermatheca. Reprinted with permission from: Van Troys M et al. Mol Biol Cell 2004; 15(10):4735-4748, ©2004 The American Society for Cell Biology.[11]

Ciboulot-knock out animals also display significant defects, in particular in the postembryonic brain.[9] Ciboulot is expressed in the larval, pupae and adult stage but enriched in the central brain during metamorphoses. In ciboulot deficient animals, specific neuronal outgrowths are incomplete, whereas ciboulot overexpression results in overgrowth of these specific neuronal structures. This indicates that ciboulot activity and its correctly balanced concentration are essential in these actin dependent migratory steps. Along the same line and conform the established biochemical activity of ciboulot in promoting actin polymerisation, doubling the amount of *D. melanogaster* profilin rescues the phenotype in ciboulot knockout organisms.

The ortholog Csp24 of the sea slug *H. crassicornis* is also expressed in neural and sensory structures for which activity or reorganization is suggested to involve actin dynamics.[38] Crow et al[12] have been able to correlate the function of this protein with the formation of intermediate memory upon conditional training using anti-sense approaches. Intriguingly, this protein is increasingly phosphorylated during conditioning in a Rho/ROCK and a cyclin dependent kinase 5 dependent manner.[39] Csp24 is the only family member thus far for which regulation by posttranslational modification has been reported.

Current data in different organisms consistently underscore an important role for the multirepeat β-thymosins in different actin based processes during development or in reproduction in line with their in vitro derived capacity to assist and modulate actin polymerization.

Acknowlegdments

S. Dhaese is a doctoral fellow of the F.W.O.-Vlaanderen. This work was supported by grants FWO-G.0157.05 (to M. Van Troys and C. Ampe), FWO-G0133.06 (to C.A), BOF 01-J04806 (to M.V.T). We thank R. Robinson and A. Aguda for sharing the coordinates of 2FF6 before release.

References

1. Lambrechts A, Van Troys M, Ampe C. The actin cytoskeleton in normal and pathological cell motility. Int J Biochem Cell Biol 2004; 6(10):1890-1909.
2. Vicente-Manzanares M, Webb DJ, Horwitz AR. Cell migration at a glance. J Cell Sci 2005; 118(21):4917-4919.
3. Wegner A. Treadmilling of actin at physiological salt concentrations. An analysis of the critical concentrations of actin filaments. J Mol Biol 1982; 161(4):607-615.
4. Carlier MF, Wiesner S, Le Clainche C et al. Actin-based motility as a self-organized system: Mechanism and reconstitution in vitro. C R Biol 2003; 326(2):161-170.
5. Bindschadler M, Osborn EA, Dewey Jr CF et al. A mechanistic model of the actin cycle. Biophys J 2004; 86(5):2720-2739.
6. dos Remedios CG, Chhabra D, Kekic M et al. Actin binding proteins: Regulation of cytoskeletal microfilaments. Physiol Rev 2003; 83(2):433-473.
7. Van Troys M, Vandekerckhove J, Ampe C. Structural modules in actin-binding proteins: Towards a new classification. Biochim Biophys Acta 1999; 1448(3):323-348.
8. Paunola E, Mattila PK, Lappalainen P. Wh2 domain: A small, versatile adapter for actin monomers. FEBS Lett 2002; 513(1):92-97.
9. Boquet I, Boujemaa R, Carlier MF et al. Ciboulot regulates actin assembly during drosophila brain metamorphosis. Cell 2000; 102(6):797-808.
10. Vandekerckhove J, Van Damme J, Vancompernolle K et al. The covalent structure of acanthamoeba actobindin. J Biol Chem 1990; 265(22):12801-12805.
11. Van Troys M, Ono K, Dewitte D et al. Tetrathymosinbeta is required for actin dynamics in caenorhabditis elegans and acts via functionally different actin-binding repeats. Mol Biol Cell 2004; 15(10):4735-4748.
12. Crow T, Redell JB, Tian LM et al. Inhibition of conditioned stimulus pathway phosphoprotein 24 expression blocks the development of intermediate-term memory in hermissenda. J Neurosci 2003; 23(8):3415-3422.
13. Herrmann D, Hatta M, Hoffmeister-Ullerich SA. Thypedin, the multi copy precursor for the hydra peptide pedin, is a beta-thymosin repeat-like domain containing protein. Mech Dev 2005; 122(11):1183-1193.

14. Crow T, Xue-Bian JJ. Identification of a 24 kda phosphoprotein associated with an intermediate stage of memory in hermissenda. J Neurosci 2000; 20(10):RC74.
15. Manuel M, Kruse M, Muller WE et al. The comparison of beta-thymosin homologues among metazoa supports an arthropod-nematode clade. J Mol Evol 2000; 51(4):378-381.
16. Dominguez R. Actin-binding proteins—a unifying hypothesis. Trends Biochem Sci 2004; 29(11):572-578.
17. Edwards J. Are beta-thymosins wh2 domains? FEBS Lett 2004; 573(1-3):231-232.
18. Vancompernolle K, Goethals M, Huet C et al. G- to f-actin modulation by a single amino acid substitution in the actin binding site of actobindin and thymosin beta 4. Embo J 1992; 11(13):4739-4746.
19. Van Troys M, Dewitte D, Goethals M et al. The actin binding site of thymosin beta 4 mapped by mutational analysis. Embo J 1996; 15(2):201-210.
20. Rossenu S, Leyman S, Dewitte D et al. A phage display-based method for determination of relative affinities of mutants. Application of the actin-binding motifs in thymosin beta 4 and the villin headpiece. J Biol Chem 2003; 278(19):16642-16650.
21. Rossenu S, Dewitte D, Vandekerckhove J et al. A phage display technique for a fast, sensitive, and systematic investigation of protein-protein interactions. J Protein Chem 1997; 16(5):499-503.
22. Czisch M, Schleicher M, Horger S et al. Conformation of thymosin beta 4 in water determined by nmr spectroscopy. Eur J Biochem 1993; 218(2):335-344.
23. Hertzog M, van Heijenoort C, Didry D et al. The beta-thymosin/wh2 domain; structural basis for the switch from inhibition to promotion of actin assembly. Cell 2004; 117(5):611-623.
24. Safer D, Sosnick T, Elzinga M. Thymosin beta 4 binds actin in an extended conformation and contacts both the barbed and pointed ends. Biochemistry 1997; 36(19):5806-5816.
25. Simenel C, Van Troys M, Vandekerckhove J et al. Structural requirements for thymosin beta4 in its contact with actin. An nmr-analysis of thymosin beta4 mutants in solution and correlation with their biological activity. Eur J Biochem 2000; 267(12):3530-3538.
26. Chereau D, Kerff F, Graceffa P et al. Actin-bound structures of wiskott-aldrich syndrome protein (wasp)-homology domain 2 and the implications for filament assembly. Proc Natl Acad Sci USA 2005; 102(46):16644-16649.
27. Domanski M, Hertzog M, Coutant J et al. Coupling of folding and binding of thymosin beta4 upon interaction with monomeric actin monitored by nuclear magnetic resonance. J Biol Chem 2004; 279(22):23637-23645.
28. Irobi E, Aguda AH, Larsson M et al. Structural basis of actin sequestration by thymosin-beta4: Implications for wh2 proteins. Embo J 2004; 23(18):3599-3608.
29. Aguda AH, Xue B, Irobi E et al. The structural basis of actin interaction with multiple wh2/beta-thymosin motif-containing proteins. Structure 2006; 14(3):469-476.
30. Huff T, Muller CS, Otto AM et al. Beta-thymosins, small acidic peptides with multiple functions. Int J Biochem Cell Biol 2001; 33(3):205-220.
31. Hertzog M, Yarmola EG, Didry D et al. Control of actin dynamics by proteins made of beta-thymosin repeats: The actobindin family. J Biol Chem 2002; 277(17):14786-14792.
32. Bubb MR, Lewis MS, Korn ED. The interaction of monomeric actin with two binding sites on acanthamoeba actobindin. J Biol Chem 1991; 266(6):3820-3826.
33. Vancompernolle K, Vandekerckhove J, Bubb MR et al. The interfaces of actin and acanthamoeba actobindin. Identification of a new actin-binding motif. J Biol Chem 1991; 266(23):15427-15431.
34. Goldschmidt-Clermont PJ, Furman MI, Wachsstock D et al. The control of actin nucleotide exchange by thymosin beta 4 and profilin. A potential regulatory mechanism for actin polymerization in cells. Mol Biol Cell 1992; 3(9):1015-1024.
35. Pantaloni D, Carlier MF. How profilin promotes actin filament assembly in the presence of thymosin beta 4. Cell 1993; 75(5):1007-1014.
36. Quinlan ME, Heuser JE, Kerkhoff E et al. Drosophila spire is an actin nucleation factor. Nature 2005; 433(7024):382-388.
37. Dedova IV, Nikolaeva OP, Safer D et al. Thymosin beta4 induces a conformational change in actin monomers. Biophys J 2006; 90(3):985-992.
38. Crow T, Xue-Bian JJ. One-trial in vitro conditioning regulates a cytoskeletal-related protein (csp24) in the conditioned stimulus pathway of hermissenda. J Neurosci 2002; 22(24):10514-10518.
39. Crow T, Xue-Bian JJ, Dash PK et al. Rho/rock and cdk5 effects on phosphorylation of a beta-thymosin repeat protein in hermissenda. Biochem Biophys Res Commun 2004; 323(2):395-401.
40. Irobi E, Burtnick L, Urosev D et al. From the first to the second domain of gelsolin: A common path on the surface of actin? FEBS Lett 2003; 552(2-3):86-90.

Chapter 8

Wasp and WAVE Family Proteins

Emanuela Frittoli, Andrea Disanza and Giorgio Scita*

Abstract

The dynamic turnover of actin filaments generates the forces driving cellular motile processes. A key factor of actin polymerization is the de novo nucleation and elongation of actin filaments, which can be catalysed by a limited number of proteins or protein complexes, the best studied of which is the Arp2/3-complex. The activity of the Arp2/3 complex is tightly regulated and controlled through signal-dependent association with nucleation promotion factors, like the WASP and WAVE family of proteins. An emerging common theme for these factors is that they act as coincident detectors of a variety of signaling pathways through the formation of large multi-molecular complexes. These complexes impose a strict spatial and temporal control on the activities of WASP and WAVE family proteins within the cells. They further contribute to fine tune Arp2/3-mediated branched actin filament elongation so as to adapt its biochemical activity to a vast array of diverse cellular functions. In this chapter we will provide an overview of the most recent finding defining the composition and mode of regulation of the WAVE-, WASP- and N/WASP-based complexes in mediating distinct actin dynamics-based cellular processes.

Introduction

The dynamic assembly of actin filaments in response to extracellular signals is at the base of a wide range of fundamental cellular processes through which living cells change shape, extend protrusions like lamellipodia and filopodia, or wrap around a particle, like in a phagocytic cup.[1-3] The bulk turnover of actin subunits is 100-200 times faster in cells than with pure actin, pointing to a complex regulation in vivo. Consistently, a large, repertoire of actin-binding proteins regulates the dynamic assembly and spatial organization of actin filaments, thus orchestrating the motile behavior of cells. Among these are proteins that: (i) promote the nucleation of actin, like the Arp2/3 complex or formins; (ii) affect the depolymerization of filaments, like the actin-depolymerizing factor (ADF/cofilin) family; (iii) associate to monomeric actin, like profilin and beta-thymosin; (iv) capping proteins. Coordination and integration of the activities of this basic set of proteins is essential to control site-directed actin polymerization in vivo.[1-3] Additionally complexity is emerging with the discovery that these proteins are, in turn, targets of various signaling pathways emanating from diverse extracellular stimuli, like those from the receptor tyrosine kinase (RTK) family.

A paradigmatic examples of these latter case is represented by a family of regulatory proteins, including WASP, N-WASP and WAVEs (1, 2, and 3). These proteins are capable to promote the assembly of three actin monomers into a trimer,[1] the rate-limiting step of actin polymerization, by directly binding and controlling the nucleation promoting activity of the

*Corresponding Author: Giorgio Scita—Istituto FIRC di Oncologia Molecolare, Via Adamello 16, 20139, Milan, Italy. Email: giorgio.scita@ifom-ieo-campus.it

Actin-Monomer-Binding Proteins, edited by Pekka Lappalainen. ©2007 Landes Bioscience and Springer Science+Business Media.

Arp2/3 complex (for review see ref. 4). Notably, this "activated" protein assembly imposes a strict topological constrain on the growth of novel actin filaments, which can only elongate in a branched fashion leading to the formation of a dendritic array of actin meshwork.[1] This is thought to be essential for the generation of a "stiff " actin network, with the actin filament fast-growing barbed ends oriented toward the extending membranes, capable of producing and supporting the forces required for propulsion, like in the lamellae of migrating cells.[5] Conversely, another mode of actin filaments elongation is catalyzed by formins, which directly bind filament barbed ends, and, independently from the Arp2/3 complex, catalyze processive growth of un-branched, linear actin filaments.[6] Here we will specifically focus on the biochemical role, the biological function, and modes of signal-dependent regulation of the WASP and WAVE family proteins, emphasizing their common and distinct properties in mediating actin polymerization both in vitro, and more importantly in in vivo relevant cellular and biological processes.

Promoting Branched Elongation of Actin Filaments through the Arp2/3 Complex

A large body of biochemical and structural biological work has defined the paradigms through which WASP and N-WASP biochemical action is exerted, by directly binding and promoting the Arp2/3-dependent actin nucleation and branched elongation activity, a key, thermodynamically un-favored step in the reaction of actin polymerization.[4,7]

The discovery of the Arp2/3 complex,[8] indeed, represented a milestone in the road toward understanding how dynamic actin turnover drives cell motility. Since dynamic assembly and disassembly of actin filaments was first put forward as a mechanism to account for ATP-dependent actin polymerization, and it was realized that cells had meshwork of actin filaments oriented to the periphery, the challenge has been the discovery of the underlying mechanisms explaining actin-based motility (reviewed in ref. 3). The identifciation of the Arp2/3 complex provided the first structural and biochemical framework to account for this.

The Arp2/3 complex was identified independently over a periods of few years following its initial isolation in *Acanthamoeba* by affinity chromatography on the actin binding protein profiling.[8] It consists of seven conserved polypeptides comprising the two actin related proteins Arp2 and Arp3, and five additional subunit commonly name according to their size (p40, p34, p21, p20 and p16), but the Human Genome Organization (HUGO) nomenclature is less ambiguous (p40 = ARPC1, p34 = ARPC2, p21 = ARPC3, p20 = ARPC4, and p16 = ARPC5). Chemical-crosslinking, yeast two hybrid analysis and genetic experiments led to a general model for the topological assembly of these seven units. This was largely confirmed and extended by recent structural analysis by cryo-electron microscopy,[9] X-ray crystallography,[10] and biochemical reconstitution experiments.[11] The Arp2/3 complex is a disk-shaped, with Arp2 and 3 sitting at the side. Both Arps display an actin-like fold with the exception of an extended region inserted in Arp3. The heart of the complex is formed by ARPC2 and ARPC4, which are closely associated, forming a cradle-like structure completed by the amino terminal extension of ARPC5. ARPC1 and ARPC3 form two basic surface patches that may serve as docking site for acidic-regions present in activators, like the WASP and WAVE family proteins.[12] Cryo-EM analysis revealed that this assembly sits at the branch junction of actin filaments, thus providing compelling evidence of its involvement in promoting branched filament elongation.[9] Despite this wealth of information the precise mechanisms of how branched elongation is achieved by activated Arp2/3 still remains disputed. Two speculative models have been proposed for subunit organization of the Arp2/3 complex at junction. The first, based largely on structural observation proposes that the complex binds to the side of a preexisting filament. Under this scenario, Arp2 and Arp3 assume an actin filament-like dimer configuration serving as a template for the initiation of daughter filaments in the barbed end direction.[13] Another conceptually different model, mainly derived from careful kinetic studies, suggested that the Arp2/3 complex induces branching and elongation at the barbed end, rather than on the side, of growing filaments.[14] Whatever the case site-directed activation of the Arp2/3 complex was proved to be the essential and critical step in the promotion of branched filaments elongation.

An important corollary and common finding of all these studies was that highly purified and isolated Arp2/3 complex has a poor capacity to initiate actin filament unless provided with ATP and activated primarily by proteins called nucleation promoting factors (NPF), and secondarily by actin filaments.[1,15] The first NPF identified was ActA from the intracellular pathogen *Listeria*.[16] This protein hijacks host cell proteins to assemble an actin filament comet tail that propels the bacterium through the cytoplasm. Proteins with related mechanism of action include the WASP/NWASP and WAVE family,[17] whose function will be described in details below.

Structural Features of WASP and N-WASP and WAVE Proteins

The first member of this family of proteins, WASP, was identified as the gene mutated in human patients with Wiskott-Aldrich syndrome (WAS). WAS is characterized by severe defects in blood clotting and in the immune system,[18] consistent with the finding that WASP is only present in hematopoietic cells. Other tissues express the ubiquitous WASP related protein, N-WASP, originally isolated from brain,[19,20] and WAVE, a WASP homologue, which was independently identified in mammals[20] and in *Dictyostelium* (named Scar).[21] WAVEs define a subfamily of proteins, which include three members, WAVE1, 2, and 3, based on distinct structural features in their N-terminal portion (Fig. 1).

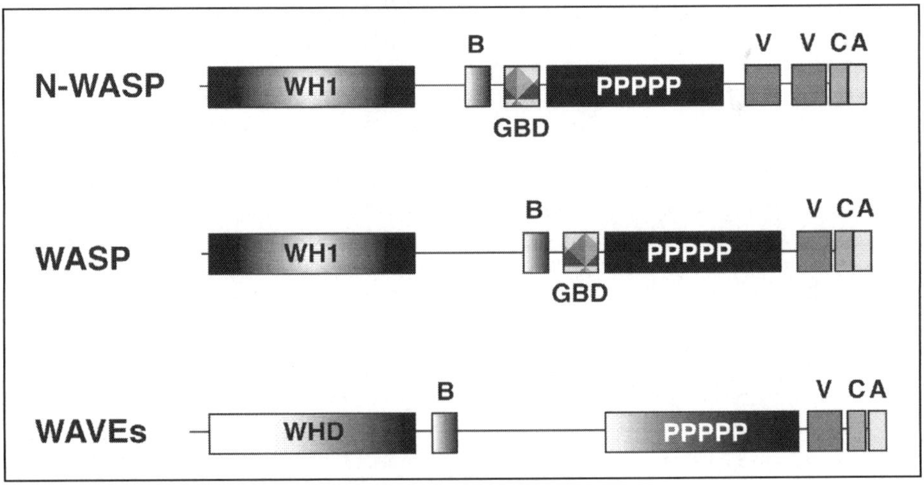

Figure 1. Modular domain organization of WASP, N-WASP and WAVE proteins. The various protein domains and regions are indicated: WH1 (WASP Homology 1) domain is conserved throughout evolution in WASP and N-WASP, but absent in the three WAVE isoforms, which instead harbor a WHD (WAVE homology domain). B indicates a stretch of basic amino acids mediating F-actin and (for WASP and N-WASP) PIP2 binding. GBD (GTPase binding domain) interacts directly with activated, GTP-loaded Cdc42. PPPPP indicates a proline-rich region containing SH3- and profilin- binding sites. V, C and A form the VCA (Verprolin homology, Cofilin homology and Acidic) modular, "output" domain responsible for initiating the growth of new actin filaments by bringing together actin monomers and the actin-nucleating Arp2/3 complex. Modified from: Stradal TE et al. Trends Cell Biol 14(6):303-311,[27] ©2004 with permission Elsevier.

All WASP and WAVE proteins share a similar modular organization with a conserved C-terminal domain and a larger and divergent N-terminal regulatory region (Fig. 1). The "output" domain consists of the VCA module, which promotes the growth of new actin filaments by linking actin monomers and the Arp2/3 complex. This occurs through a Verprolin homology domain (V, also called a WASP homology, WH2 domain) and a C-terminal acidic (A) region binding to an actin monomer[22] and Arp2/3 complex, respectively.[23] A conserved sequence (C), named Cofilin homology domain acts together with the V and A region,[24-26] driving the conformational changes necessary to stimulate nucleation. The analysis of the catalytic properties of the VCA module of WASP, N-WASP, and WAVE1 revealed that the isolated domains display unique kinetics of actin assembly reflecting slight but significant structural differences.[27] Thus, despite nearly equal affinity of the various VCA domains for binding to Arp2-3, the rate of actin nucleation can significantly differ, suggesting that a fine tuning in the catalytic process of activation of Arp2/3 by VCA domains may result in different dynamic rates of actin assembly ultimately affecting the architecture of actin networks produced by the different NPFs.

The overall conservation among all the WASP and WAVE family members within the C-terminal output domain, contrasts with the divergence of their N-terminal portion (Fig. 1). WASP and N-WASP display a common modular organization, including the WH1/EVH1 (WASP homology/ENA VASP homology), a basic region, a GBD/CRIB (GTPases binding domain/Cdc42 and Rac Interactive Binding Domain) and a proline rich region.

The WH1 domain encompasses the first 150 amino acids.[28,29] Its functional significance is underscored by the observation that most of the identified mutations leading to the disease WAS mapped within this domain.[18] Notably, the N-WASP WH1 associates primarily and tightly with a 25 motif residues of WIP,[30] a member of a family of regulators of N-WASP mediated actin polymerization,[31] including WICH and CR16.[32,33] Since mutations in WH1 found in WAS patients may lead to disruption of the interaction with WIP family members, it is likely the WIP-WASP interaction is essential for the proper functional activity of the complex.

The basic, GBD/CRIB domain, and the proline rich region have all been implicated in regulating WASP and N-WASP function by exerting or reverting auto-inhibitory interactions that block the activity of the "output" VCA domain.[26,34] Binding to upstream activators relieves these inhibitory interactions.[26,34] For example, activated GTP-loaded Cdc42 associates with the GBD/CRIB motif, PIP2 (phosphatidylnositol 4, 5 phosphates) and F-actin bind to the basic region, and a plethora of SH3-containing proteins,[4,35,36] and profilin[37] associate to the polyproline motives in the proline rich region (see below).

The N-terminal region of WAVE proteins is less characterized. At variance from WASP and N-WASP, no GBD/CRIB motif and no WH1 domain are present in WAVEs.[4] The lack of a surface directly linking WAVEs to Rho GTPases indicated that these proteins are regulated in a different fashion with respect to WASP and N-WASP. Consistently, WAVEs N-terminal region, defined as WHD (WAVE homology domain), is highly conserved among WAVE1,2 and 3 and across evolution. The WHD domain does not associates to WIP family. It was, instead, found to mediate a direct binding to Abi1,[38] a scaffolding molecule originally identified as an interactor of the nonreceptor tyrosine kinase, Abl,[39] and involved in Ras to Rac signaling.[40] This interaction drives the assembly of a WAVE-based macromolecular complex, which, in turn, mediates a direct association to activated Rac, and regulates in vitro WAVE2 Arp2/3-dependent actin polymerization.[38,41,42] Finally, similar to WASP and N-WASP, WAVE contains a basic motif and a proline rich region. The latter mediates the association with the insulin receptor substrate, IRSp53, which has been implicated in physically linking WAVE2 to Rac.[43] The former was recently shown to mediate binding with PIP3 (phosphatidylnositol 3, 4, 5 phosphates), which was proposed to spatially restrict WAVE localization at the leading edge of membrane protrusions.[44]

WASP, N-WASP and WAVE are regulated in vivo through protein:protein interactions functioning in distinct actin dynamics-based processes in cells and organisms.

N-WASP Mediated Actin Polymerization Events

As mentioned, WASP and N-WASP associate to a plethora of binding partners most of which are canonical signaling proteins.[27] These signaling intermediates are thought to act in a concerted manner, imposing a strict control on the extent, duration and location of WASP and N-WASP activity (Fig. 2).

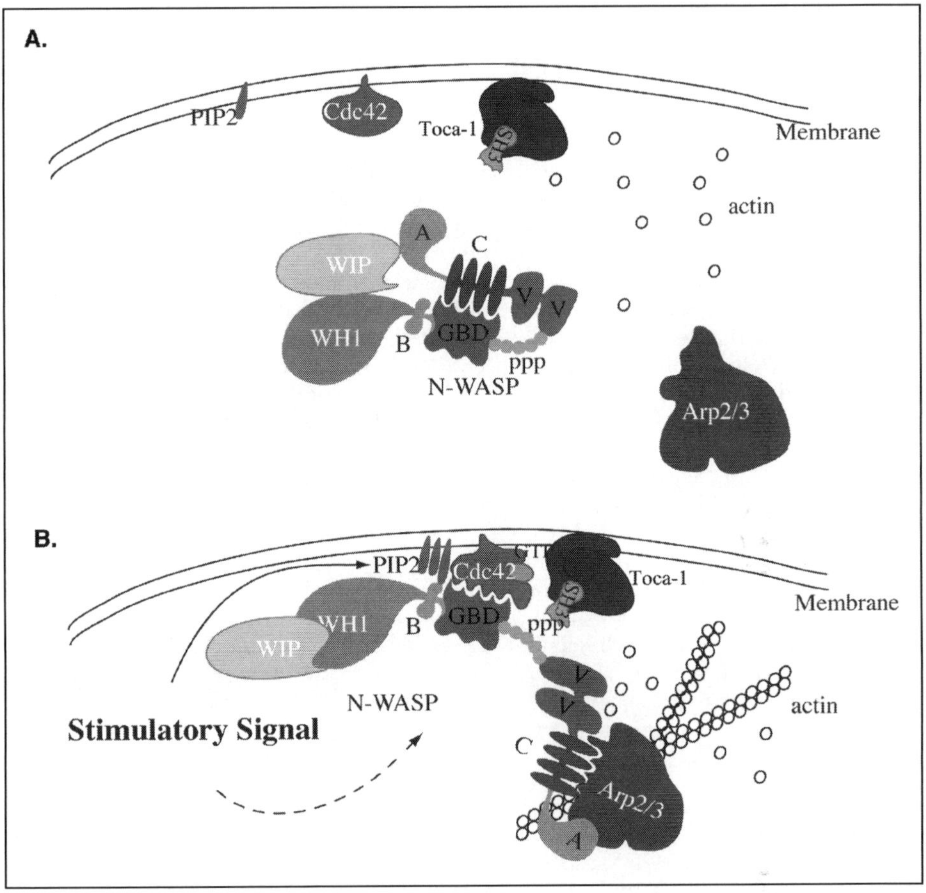

Figure 2. Model of activation of N-WASP by multiple signals. In un-stimulated cells, WASP and N-WASP are locked in an auto-inhibited state, which is aided by the association with WIP family proteins. A) A number of stimulatory events, like those emanating from of a variety of activated membrane receptors, may lead to increase and localized PIP2 production and Cdc42 activation. Cooperative binding of PIP2 and Cdc42, to the B region and the GBD domain of N-WASP, respectively, causes a conformational change, resulting in release of the C-terminus and enabling the activation of Arp2/3 with the ensuing formation of branched actin filaments. B) PIP2 may also uncap filament barbed ends, favouring actin polymerization. Additionally interaction with a number of SH3-domain-containing proteins, including Toca-1, and post-translational modifications, like phosphorylation, cooperate in modulating the activity of N-WASP. WH1, WASP homology domain; B, basic region; PPP, proline-rich region; V, Verprolin Homology domain; C, Cofilin Homology domain; A, acidic region; SH3, Src homology 3 domain. Modified from: Stradal TE et al. Trends Cell Biol 14(6):303-311,[27] ©2004 with permission Elsevier.

Until recently, it was widely accepted that N-WASP existed as an auto-inhibited monomer, reflecting its in vitro biochemical properties. The binding of N-WASP to activators, particularly Cdc42 and PIP2, would relieve this inhibitory interaction.[34] This provided a simple and elegant model accounting for signalling-dependent modulation of N-WASP-mediated actin polymerization. Recently, however, biochemical approaches indicated that N-WASP and WASP are mainly bound to WIP family members, like the brain specific CR16[45] or WIP.[46,47] More importantly, WIP binding to N-WASP was shown to stabilize, at least in vitro, its inactive conformation, supporting the notion that the WIP-N-WASP complex is the relevant auto-inhibited signaling unit, whose modes of regulation may differs from that of isolated N-WASP.[47] Consequently, Ho and colleagues, using Xenopus egg extracts, identified a novel protein, Toca-1 (Transducer of Cdc42-dependent actin assembly), required for full, Cdc42-induced activation of the WIP-WASP complex.[47] Toca-1 has the structural features of a typical effector/adaptor molecule being capable of directly associating to activated Cdc42, via its HR1- and to N-WASP via its SH3-domain. The addition of Toca-1 to native WIP-WASP complex, but not to isolated N-WASP was necessary for Cdc42 and PIP2-mediated activation. Consistently, Toca-1 was also shown to be required for PMA-induced vesicle rocketing, which strictly depends on WASP or N-WASP.[48] This finding, while highlighting a level of unexpected complexity on the physiological regulation of WASP and N-WASP, also raised a number of additional issues. Biochemically, whether the binding of Toca-1 to WASP displaces WIP or changes the overall conformation of the WIP-N-WASP complex, exposing its VCA domain, has not been defined. Similarly, the overall topology of a putative Cdc42/Toca-1/WIP/WASP complex remains to be elucidated. More biological complexity also comes from the observation that Toca-1 is a member of a highly conserved three-gene family comprising Cip4, FBP17 and Toca-1. It is unclear, yet, whether these proteins have overlapping functions with respect to their presumed functional role on WASP. A recent intriguing observation indicated that Cip4, FBP17 and Toca-1 all share a conserved N-terminal domain which extend with respect to the previously identified, but functionally uncharacterized FCH (Fes/Cip4 Homology) domain, to include coil-coiled helices resembling overall a specialized BAR (Bin-Amphyphisin-RSV) domain, named F-BAR.[49] BAR domain are generally found in endocytic proteins (for review see ref. 50), and are capable of binding lipid bi-layers, sensing their curvature and inducing their tubulation. Accordingly Cip4, FBP17 and Toca-1 were found to lead to formation of lipid tubulation both in vitro and in vivo and to act in concert with Dynamin, a GTPase mediating constriction and fission of lipid vesicles and tubules during the process of endocytosis.[49] This together with the finding that Toca-1 can control also the WIP-WASP complex provide a direct molecular framework connecting WASP and N-WASP-mediated actin assembly with membrane traffic and endocytosis. Finally, the importance of WIP family protein in regulation of WASP activity found further support in the observation that the complex appears to be conserved also in lower organism, like the nematodes.[51] In this organism, individual ablation by RNAi of either WIP-1 or WSP-1 (the *C. elegans* homologue of WASP and N-WASP, respectively) leads to embryonic lethality due to ventral closure defects.[51] Furthermore and more importantly, biochemical and genetic interaction between WIP-1 and WSP-1 could be shown.

All these studies highlighted an important concept whereby WASP and N-WASP functional activity strictly depends on its engagement into distinct macromolecular complexes which likely act by functionally specifying N-WASP activity in different biological processes. This may also account for the large and ever growing number of N-WASP binding partners. As a result WASP and N-WASP were demonstrated to play a versatile role in actin assembly events at the plasma membrane or at vesicles. For instance, N-WASP null fibroblasts fail to support the movement of endosomal vesicles evoked by increased PIP2 or phosphotyrosine levels, a phenotype which can be restored by re-expression of either N-WASP or WASP.[48] WASP is essential for the formation of podosomes, specialized actin-rich adhesive structure[52] and several lines of evidence link WASP and N-WASP to receptor-mediated endocytosis. WASP deficiency is accompanied by impaired endocytosis of the T-cell antigen receptor.[53] Moreover, WASP and

N-WASP are known to interact with a variety of endocytic proteins, like syndapin, intersectin, cortactin and indirectly with dynamin.[54-56] By employing two-colours TIRF microscopy direct visualisation of recruitment of actin and N-WASP accompanying internalisation of single clathrin-coated pits could be achieved.[57,58] Finally, endocytic defects (reduced rate of initial internalisation[59] and increased levels of cell-surface membrane receptors)[60] were observed in cells ablated for N-WASP expression by genetic deletion[59] or by RNA-interference.[60]

The endocytic function of WASP and N-WASP are evolutionary conserved. The only WASP/Scar-family member in yeast, Las17 (also termed Bee1p) is an important constituent of the cortical actin patches, crucial for endocytosis and polarity. Like in vertebrates N-WASP, Las17 binds to verprolin (the yeast WIP-family homologue),[61] and to the yeast Toca-1 homologue, BZZ1p. Moreover, it is transiently recruited to endocytic clathrin patches together with a variety of additional actin regulatory proteins, which represent constitutive and essential components of the endocytic machinery.[62,63]

Another level of signal-integration and N-WASP modulation is achieved by post-translational modifications. Both, WASP and N-WASP are phosphorylated on serine and tyrosine residues. First, phosphorylation of a serine located within the VCA domain increases its affinity for Arp2/3 complex.[64] Second, cytoplasmic tyrosine kinases can phosphorylate WASP and N-WASP on a central tyrosine,[65,66] which is only accessible to both kinases and phosphatases in the activated state of N-WASP, i.e., not in the auto-inhibited conformation, and thus suggesting an additional mechanism for N-WASP regulation through molecular memory.[67]

WAVE-Complexes Acts as Signaling Machineries in Cellular Protrusions

There are three structurally conserved WAVE proteins (WAVE1, 2, and 3) present in mammals. They all display constitutive activity, and mediate actin remodelling by indirectly associating to Rac. The insulin receptor substrate IRSp53 was identified as the first potential link between WAVE2 and Rac.[43] However, while a complex including Rac-IRSp53-WAVE2 could be reconstituted in vitro, it did not significantly affect the nucleating promoting function of WAVE2.[43] Alternative and more complex mechanisms to control WAVEs via Rac have been recently proposed and stimulated a controversial debate.[38,68] WAVE1, from brain lysates, was found to be associated with three other proteins: Nap1,[69] PIR121/Sra-1, identified as a Rac effector[70] and HSPC300.[68] Abi-2, an Abl interactor,[71] was also found part of this complex. This complex (hereafter referred to as WAVE-complex) was unable to stimulate actin polymerization in in vitro assays. Addition of active Rac induced the disassembly of the inhibitory Nap1-PIR121 sub-complex from the active WAVE1-HSPC300 unit[68] (Fig. 3A). More recently, in vivo and in vitro reconstitution experiments challenged this attractive inhibitory model arguing, instead, for a positive mode of regulation exerted by the assembly of WAVE into a Abi1-Nap1-PIR121 complex. Moreover, the addition of activated Rac did not disrupt the complex[38,41] (Fig. 3B). Notwithstanding these differences, it is generally accepted that the great majority of cellular WAVE proteins[72] are engaged in stable complexes containing only HSPC300, WAVE, Abi1, Nap1 and PIR121, which assemble in a precise order such that Abi1 and Nap1 sit at the core of the complex linking WAVE and HSPC300 on one hand, and the Rac effector, PIR121/Sra-1, on the other.[42] This topology is highly conserved from *Drosophila* to *C. elegans* and *Dictyostelium*, but also in plants.[73] Additionally, in vivo, all the components of WAVE-complex are rapidly and simultaneously relocalized to leading edges of extending membrane protrusion in response to Rac activation.[41] This suggests that the complex 'moves" as a whole. Consistently, removal of any of the subunits disrupts the localization of the others.[38,41] The development of FRET probes, however, exploiting the close proximity of Nap1 to Abi1 and WAVE2 would be needed to unequivocally validate this assumption. Furthermore, removal of single subunits results in downregulation of the other subunits, probably by protein degradation.[38,41,74,75]

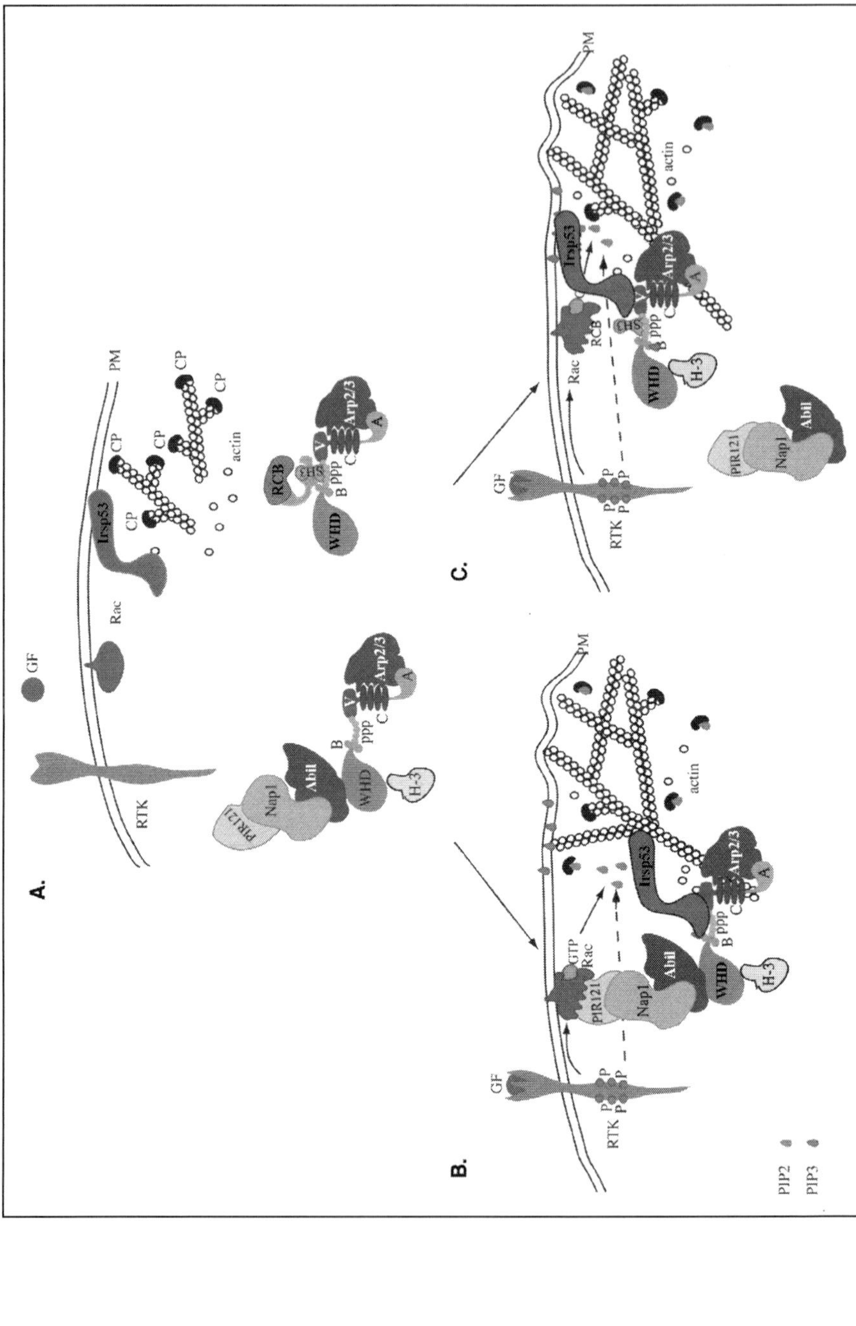

Figure 3. Please see figure legend on the next page.

Figure 3, viewed on previous page. Models for the activation of the WAVE-containing complexes. A) In resting, nonmotile cells, WAVE proteins are assembled in a stable complex including Abi1, Nap1, PIR/Sra1 and HSPC300 (H-3). The soluble and native complex isolated from lysate appear to be unable to bind Arp2/3 and stimulate actin polymerization.[42,68] However, reconstitution of the same complex using purified components resulted, instead, in an active WAVE-complex capable of promoting Arp2/3-mediated branched elongation.[38] The reason for this discrepancy remains to be investigated (see text for details). Whatever the case actin polymerization induced by WAVE must be kept inhibited in the cytoplasm. Under these conditions, filaments exist but barbed ends are mostly capped by capping proteins (CP),[86] thereby preventing unwanted actin polymerization. B) Following Rac stimulation, by activation of membrane receptors, for instance, coordination between uncapping of filaments and de novo actin nucleation occurs. This is achieved through the site-restricted production of PIP2,[86] which negatively regulates the activity of capping proteins, and PIP3, which may aid in spatially restricting the activity of WAVE to protruding leading edges. Rac activation following RTK stimulation is proposed to exert a activation of the WAVE-complex though two distinct modalities that are at least, in part, still highly debated: B) according to this model the WAVE complex, which remain stables even upon direct binding to GTP-loaded Rac, relocalizes to the leading edges. The integrity of the complex is necessary for Rac-mediated relocalization.[38,41,74] C) An alternative mode of action has also been proposed according to which the assembly of a WAVE1-HSPC300-Nap1-PIR121 complex would result in trans-inhibition of WAVE1 activity, which could be relieved following binding to activated Rac by disruption of the complex in distinct subunits.[68] However, no additional evidence has been provided supporting the Rac-induced disruption of the WAVE-complex. Thus, a possible reconciling view between these models is that while the WAVE complex is highly stable, additional events may intervene in either negative regulating its activity in the cytoplasm or promoting its activation at the plasma membranes. WHD, WAVE Homology domain; B, basic region; PPP, proline-rich region; V, Verprolin Homology domain; C, Cofilin Homology domain; A, acidic region; SH3, Src homology 3 domain.

The apparent discrepancies between two possible mechanisms of activation, dissociation versus stable activation, may reflect different experimental set-ups. Alternatively, multiple "activating" switches may contribute to the regulation of WAVE-complex activity in vivo (Fig. 3). The basic region of WAVE, for instance, was recently shown to bind PIP3.[44] PIP3 was concluded to be of major importance in recruiting WAVE2 to leading edges of membrane protrusion even in the presence of a dominant negative Rac or of Latrunculin A, a G-actin sequestering compound. Furthermore, a WAVE2 mutant impaired in PIP3 binding inhibited the formation of lamellipodia.[44] Despite the fact that it was not tested whether the actin polymerization activity or the association of WAVE2 to its binding partners was affected by PIP3, it is reasonable to assume that the localized and restricted production of this phospholipid may participate in the proper recruitment/regulation of the WAVE.

Phosphorylation may also regulate WAVE2 activity. WAVE2 was shown to be a substrate of the nonreceptor tyrosine kinase, Abl.[76] Notably, Abi1 is required to couple Abl and WAVE2. Whether this modification activates WAVE2-mediated actin polymerization or it is necessary for its proper localization remains to be demonstrated.

Finally, additional WAVE-complex binding partners may also participate in optimally modulating either the activity or the localization underneath the plasma membranes. These interactors would be expected to act either transiently in association with the WAVE complex or in subcellular compartments, like the plasma membranes. Thus, they may have escape purification procedures, which inevitably relay on the solubility of all the components. An intriguing candidate to serve this role may be represented by IRSp53, which is clearly not a core component of the WAVE-complex, but may become associated exclusively at the leading edges, where it appears to be highly enriched[77] (Fig. 3). Similarly, WAVE2 was also shown to bind to Rac-specific GTPase activating protein (GAP), WRP, which may act as a signal terminator for Rac.[78] Thus, the WAVE2-based complex may act as bona fide macromolecular detector with a built-in intrinsic mode of positive and negative regulation, which ultimately determines its final biological outcome. More experimental work is likely to shed some light in this direction and hopefully lead to a scenario capable of reconciling all the experimental observations made providing a more compelling model of how WAVE proteins are regulated.

The molecular complexity of the WAVE-complex may also account for the discrepancy among the biological phenotypes observed following individual removal of single components of the complex, in different model organisms. Individual removal of WAVE2, Abi1, Nap1/Kette or PIR121/Sra-1, in either mammalian or *Drosophila* cells, impairs the formations of membrane protrusions.[38,41,74,75] Similarly, in plants, removal of the WAVE plant homologue,[79] results in distorted trichomes, which is phenocopied by removal of the plant homologues of PIR121 (PIRP/PIROGI/AtPIR) or Nap1 (NAPP/NAPP/GNARLED/ AtNAP).[80] Surprisingly, instead, knock-out of PirA, the *Dictyostelium* homologue of PIR121,[81] resulted in uncontrolled, elevated actin polymerization and disregulated membrane protrusions, which was genetically dependent on WAVE, suggestive of an inhibitory role of PIR121 on WAVE, despite WAVE was barely detectable in these cells.[81] Conversely, removal of the Nap1 homologues in *Dictyostelium*, reduces pseudopodia formations[82] and resembles the WAVE/Scar null phenotype[21] in agreement with a positive role of Nap1 in controlling WAVE activity. On the first glimpse, these results appear contradictory. *Dictyostelium*, however, are professional movers, displaying an amoeboid-type of motility, which may only in part be mediated by WAVE-complex based actin assembly or whose WAVE-complex components may have acquired additional functions and biochemical properties. Notably, behaviour of the PirA knock-out *Dictyostelium* cells is reminiscent of the phenotype reported for WASP deficiency,[83] suggesting a connection of WAVE and WASP-mediated pathways. Biochemical characterization of the *Dictyostelium* WAVE-, and WASP- based complexes would be required to shed some lights on this issue. Similarly complex, removal of Kette, the *Drosophila* homologue of Nap1, led to a disorganized actin distribution, which was partly WAVE-dependent, suggesting an antagonizing function exerted by Nap1/Kette on WAVE.[84] The interpretation of these results is, once again, difficult especially since Nap1/Kette in *Drosophila* genetically interacts with WASP. This suggests the possibility that members of the WAVE-complex may form biochemical distinct and functional diverse macromolecular signaling units. This concept is corroborated by recent reports, demonstrating that Abi1 also plays a critical role in regulating N-WASP-dependent actin polymerisation. Abi1 was found able to bind and potently activate N-WASP in a Cdc42-cooperative fashion, through its C-terminal SH3-domain, suggesting the unexpected possibility that Abi1 could simultaneously link WAVE and N-WASP. Notably, however, a complex containing both WAVE and N-WASP could not be detected in vivo. Furthermore, molecular genetic approaches demonstrated that, in mammals, Abi1 and WAVE, but not N-WASP, are essential for Rac-dependent membrane protrusion. Conversely, Abi1 and N-WASP, but not WAVE regulated actin-based vesicular trafficking, as well as cell-surface distribution and endocytosis of receptors such as epidermal growth factor- and transferring receptor.[59,60] Similarly, in *Drosophila*, genetic approaches indicated that Abi1 and N-WASP are required for bristles development.[85] Moreover, activation of Abi1 resulted in the formation of ectopic bristle, which was dependent on WASP, but not on WAVE.[85]

Thus, a scenario is emerging whereby the regulated exchange between different complexes and diverse sub-cellular localization of individual subunits (like Abi1) (e.g., WAVE versus N-WASP), dictates their functional specificity in actin dynamics-based processes.

Conclusions and Perspective

The wealth of studies focusing on the molecular mechanisms of how elongation of actin filaments is achieved in vitro and in vivo has revealed a remarkable level of complexity, which was unexpected based on inevitably reductionistic approaches of biochemical reconstitution. The family of WASP, NWASP and WAVE despite sharing basic and common molecular mechanisms to activate the Arp2/3 complex and drive branched elongation of actin filaments, has clearly acquired in the course of the evolution a great variety of different and distinct structural features. This has enable this family protein to adapt and fine tuning their biochemical output to the various and specialized needs that processes as diverse as cell migration, and intracellular trafficking required. Plasticity in the biochemical response has therefore been achieved through a continuous process of subtle diversification of the regulatory moiety of these proteins and their interactors, while maintaining an impressive level of conservation in the domains directly involved in actin polymerization and elongation. The specific and distinct roles of the various members of this family are only beginning to be appreciated in complex organisms. This will surely represent one of the major challenges ahead requiring the precise analysis of the proteins complexes and signaling dependent post-translational modifications that act on WASP, N-WASP, WAVE1, WAVE2, WAVE3 in the ever increasing number of fundamental biological processes they are involved. The advent of more sophisticated imaging techniques, coupled with traditional biochemical and genetic approaches is likely to change the scenario of how, we think, actin dynamics is regulated.

Acknowledgements

This work was in part supported by grants from AIRC (Associazione Italiana Ricerca sul Cancro) and AIRC regione Lombardia to G. Scita, from Human Science Frontier Program to G. Scita (grant # RGP0072/2003-C), and from European Community (VI Framework) to G. Scita and A. Disanza.

References

1. Pollard TD, Borisy GG. Cellular motility driven by assembly and disassembly of actin filaments. Cell 2003; 112(4):453-465.
2. Small JV, Stradal T, Vignal E et al. The lamellipodium: Where motiliy begins. Trends Cell Biol 2002; 12(3):112-120.
3. Pantaloni D, Le Clainche C, Carlier MF. Mechanism of actin-based motility. Science 2001; 292(5521):1502-1506.
4. Takenawa T, Miki H. WASP and WAVE family proteins: Key molecules for rapid rearrangement of cortical actin filaments and cell movement. J Cell Sci 2001; 114(Pt 10):1801-1809.
5. Svitkina TM, Borisy GG. Progress in protrusion: The tell-tale scar. Trends Biochem Sci 1999; 24(11):432-436.
6. Zigmond SH. Formin-induced nucleation of actin filaments. Curr Opin Cell Biol 2004; 16(1):99-105.
7. Prehoda KE, Lim WA. How signaling proteins integrate multiple inputs: A comparison of N-WASP and Cdk2. Curr Opin Cell Biol 2002; 14(2):149-154.
8. Machesky LM, Atkinson SJ, Ampe C et al. Purification of a cortical complex containing two unconventional actins from Acanthamoeba by affinity chromatography on profilin-agarose. J Cell Biol 1994; 127(1):107-115.
9. Volkmann N, Amann KJ, Stoilova-McPhie S et al. Structure of Arp2/3 complex in its activated state and in actin filament branch junctions. Science 2001; 293(5539):2456-2459.
10. Robinson RC, Turbedsky K, Kaiser DA et al. Crystal structure of Arp2/3 complex. Science 2001; 294(5547):1679-1684.
11. Gournier H, Goley ED, Niederstrasser H et al. Reconstitution of human Arp2/3 complex reveals critical roles of individual subunits in complex structure and activity. Mol Cell 2001; 8(5):1041-1052.
12. Winder SJ. Structural insights into actin-binding, branching and bundling proteins. Curr Opin Cell Biol 2003; 15(1):14-22.
13. Amann KJ, Pollard TD. Direct real-time observation of actin filament branching mediated by Arp2/3 complex using total internal reflection fluorescence microscopy. Proc Natl Acad Sci USA 2001; 98(26):15009-15013.
14. Pantaloni D, Boujemaa R, Didry D et al. The Arp2/3 complex branches filament barbed ends: Functional antagonism with capping proteins. Nat Cell Biol 2000; 2(7):385-391.
15. Carlier MF, Wiesner S, Le Clainche C et al. Actin-based motility as a self-organized system: Mechanism and reconstitution in vitro. C R Biol 2003; 326(2):161-170.
16. Welch MD, Rosenblatt J, Skoble J et al. Interaction of human Arp2/3 complex and the Listeria monocytogenes ActA protein in actin filament nucleation. Science 1998; 281(5373):105-108.

17. Stradal TE, Scita G. Protein complexes regulating Arp2/3-mediated actin assembly. Curr Opin Cell Biol 2005.
18. Imai K, Nonoyama S, Ochs HD. WASP (Wiskott-Aldrich syndrome protein) gene mutations and phenotype. Curr Opin Allergy Clin Immunol 2003; 3(6):427-436.
19. Miki H, Miura K, Matuoka K et al. Association of Ash/Grb-2 with dynamin through the Src homology 3 domain. J Biol Chem 1994; 269(8):5489-5492.
20. Miki H, Suetsugu S, Takenawa T. WAVE, a novel WASP-family protein involved in actin reorganization induced by Rac. EMBO J 1998; 17(23):6932-6941.
21. Bear JE, Rawls JF, Saxe IIIrd CL. SCAR, a WASP-related protein, isolated as a suppressor of receptor defects in late Dictyostelium development. J Cell Biol 1998; 142(5):1325-1335.
22. Miki H, Takenawa T. Direct binding of the verprolin-homology domain in N-WASP to actin is essential for cytoskeletal reorganization. Biochem Biophys Res Commun 1998; 243(1):73-78.
23. Machesky LM, Insall RH. Scar1 and the related Wiskott-Aldrich syndrome protein, WASP, regulate the actin cytoskeleton through the Arp2/3 complex. Curr Biol 1998; 8(25):1347-1356.
24. Marchand JB, Kaiser DA, Pollard TD et al. Interaction of WASP/Scar proteins with actin and vertebrate Arp2/3 complex. Nat Cell Biol 2001; 3(1):76-82.
25. Panchal SC, Kaiser DA, Torres E et al. A conserved amphipathic helix in WASP/Scar proteins is essential for activation of Arp2/3 complex. Nat Struct Biol 2003; 10(8):591-598.
26. Kim AS, Kakalis LT, Abdul-Manan N et al. Autoinhibition and activation mechanisms of the Wiskott-Aldrich syndrome protein. Nature 2000; 404(6774):151-158.
27. Stradal TE, Rottner K, Disanza A et al. Regulation of actin dynamics by WASP and WAVE family proteins. Trends Cell Biol 2004; 14(6):303-311.
28. Symons M, Derry JM, Karlak B et al. Wiskott-Aldrich syndrome protein, a novel effector for the GTPase CDC42Hs, is implicated in actin polymerization. Cell 1996; 84(5):723-734.
29. Miki H, Miura K, Takenawa T. N-WASP, a novel actin-depolymerizing protein, regulates the cortical cytoskeletal rearrangement in a PIP2-dependent manner downstream of tyrosine kinases. EMBO J 1996; 15(19):5326-5335.
30. Volkman BF, Prehoda KE, Scott JA et al. Structure of the N-WASP EVH1 domain-WIP complex: Insight into the molecular basis of Wiskott-Aldrich Syndrome. Cell 2002; 111(4):565-576.
31. Ramesh N, Anton IM, Hartwig JH et al. WIP, a protein associated with wiskott-aldrich syndrome protein, induces actin polymerization and redistribution in lymphoid cells. Proc Natl Acad Sci USA 1997; 94(26):14671-14676.
32. Kato M, Miki H, Kurita S et al. WICH, a novel verprolin homology domain-containing protein that functions cooperatively with N-WASP in actin-microspike formation. Biochem Biophys Res Commun 2002; 291(1):41-47.
33. Zettl M, Way M. The WH1 and EVH1 domains of WASP and Ena/VASP family members bind distinct sequence motifs. Curr Biol 2002; 12(18):1617-1622.
34. Prehoda KE, Scott JA, Dyche Mullins R et al. Integration of multiple signals through cooperative regulation of the N- WASP-Arp2/3 complex. Science 2000; 290(5492):801-806.
35. Rohatgi R, Ho HY, Kirschner MW. Mechanism of N-WASP activation by CDC42 and phosphatidylinositol 4, 5- bisphosphate. J Cell Biol 2000; 150(6):1299-1310.
36. Suetsugu S, Miki H, Yamaguchi H et al. Enhancement of branching efficiency by the actin filament-binding activity of N-WASP/WAVE2. J Cell Sci 2001; 114(Pt 24):4533-4542.
37. Yang C, Huang M, DeBiasio J et al. Profilin enhances Cdc42-induced nucleation of actin polymerization. J Cell Biol 2000; 150(5):1001-1012.
38. Innocenti M, Zucconi A, Disanza A et al. Abi1 is essential for the formation and activation of a WAVE2 signalling complex. Nat Cell Biol 2004; 6(4):319-327.
39. Shi Y, Alin K, Goff SP. Abl-interactor-1, a novel SH3 protein binding to the carboxy-terminal portion of the Abl protein, suppresses v-abl transforming activity. Genes Dev 1995; 9(21):2583-2597.
40. Innocenti M, Frittoli E, Ponzanelli I et al. Phosphoinositide 3-kinase activates Rac by entering in a complex with Eps8, Abi1, and Sos-1. J Cell Biol 2003; 160(1):17-23.
41. Steffen A, Rottner K, Ehinger J et al. Sra-1 and Nap1 link Rac to actin assembly driving lamellipodia formation. EMBO J 2004; 23(4):749-759.
42. Gautreau A, Ho HY, Li J et al. Purification and architecture of the ubiquitous Wave complex. Proc Natl Acad Sci USA 2004; 101(13):4379-4383.
43. Miki H, Yamaguchi H, Suetsugu S et al. IRSp53 is an essential intermediate between Rac and WAVE in the regulation of membrane ruffling. Nature 2000; 408(6813):732-735.
44. Oikawa T, Yamaguchi H, Itoh T et al. PtdIns(3,4,5)P3 binding is necessary for WAVE2-induced formation of lamellipodia. Nat Cell Biol 2004; 6(5):420-426.

45. Ho HY, Rohatgi R, Ma L et al. CR16 forms a complex with N-WASP in brain and is a novel member of a conserved proline-rich actin-binding protein family. Proc Natl Acad Sci USA 2001; 98(20):11306-11311.
46. Anton IM, de la Fuente MA, Sims TN et al. WIP deficiency reveals a differential role for WIP and the actin cytoskeleton in T and B cell activation. Immunity 2002; 16(2):193-204.
47. Ho HY, Rohatgi R, Lebensohn AM et al. Toca-1 mediates Cdc42-dependent actin nucleation by activating the N-WASP-WIP complex. Cell 2004; 118(2):203-216.
48. Benesch S, Lommel S, Steffen A et al. Phosphatidylinositol 4,5-biphosphate (PIP2)-induced vesicle movement depends on N-WASP and involves Nck, WIP, and Grb2. J Biol Chem 2002; 277(40):37771-37776.
49. Itoh T, Erdmann KS, Roux A et al. Dynamin and the actin cytoskeleton cooperatively regulate plasma membrane invagination by BAR and F-BAR proteins. Dev Cell 2005; 9(6):791-804.
50. Peter BJ, Kent HM, Mills IG et al. BAR domains as sensors of membrane curvature: The amphiphysin BAR structure. Science 2004; 303(5657):495-499.
51. Sawa M, Takenawa T. Caenorhabditis elegans WASP-interacting protein homologue WIP-1 is involved in morphogenesis through maintenance of WSP-1 protein levels. Biochem Biophys Res Commun 2006; 340(2):709-717.
52. Jones GE, Zicha D, Dunn GA et al. Restoration of podosomes and chemotaxis in Wiskott-Aldrich syndrome macrophages following induced expression of WASp. Int J Biochem Cell Biol 2002; 34(7):806-815.
53. McGavin MK, Badour K, Hardy LA et al. The intersectin 2 adaptor links Wiskott Aldrich Syndrome protein (WASp)-mediated actin polymerization to T cell antigen receptor endocytosis. J Exp Med 2001; 194(12):1777-1787.
54. Kessels MM, Qualmann B. Syndapins integrate N-WASP in receptor-mediated endocytosis. EMBO J 2002; 21(22):6083-6094.
55. Hussain NK, Wasiak S, Lamarche-Vare N et al. The brain specific longform of intersectin functions as a guanine nucleotide exchange factor for cdc42. Mol Biol Cell 2001; 11:611-620.
56. Engqvist-Goldstein AE, Drubin DG. Actin assembly and endocytosis: From yeast to mammals. Annu Rev Cell Dev Biol 2003; 19:287-332.
57. Merrifield CJ, Feldman ME, Wan L et al. Imaging actin and dynamin recruitment during invagination of single clathrin-coated pits. Nat Cell Biol 2002; 4(9):691-698.
58. Merrifield CJ. Seeing is believing: Imaging actin dynamics at single sites of endocytosis. Trends Cell Biol 2004; 14(7):352-358.
59. Benesch S, Polo S, Lai FP et al. N-WASP deficiency impairs EGF internalization and actin assembly at clathrin-coated pits. J Cell Sci 2005; 118(Pt 14):3103-3115.
60. Innocenti M, Gerboth S, Rottner K et al. Abi1 regulates the activity of N-WASP and WAVE in distinct actin-based processes. Nat Cell Biol 2005.
61. Naqvi SN, Zahn R, Mitchell DA et al. The WASp homologue Las17p functions with the WIP homologue End5p/verprolin and is essential for endocytosis in yeast. Curr Biol 1998; 8(17):959-962.
62. Lechler T, Jonsdottir GA, Klee SK et al. A two-tiered mechanism by which Cdc42 controls the localization and activation of an Arp2/3-activating motor complex in yeast. J Cell Biol 2001; 155(2):261-270.
63. Soulard A, Lechler T, Spiridonov V et al. Saccharomyces cerevisiae Bzz1p is implicated with type I myosins in actin patch polarization and is able to recruit actin-polymerizing machinery in vitro. Mol Cell Biol 2002; 22(22):7889-7906.
64. Cory GO, Cramer R, Blanchoin L et al. Phosphorylation of the WASP-VCA domain increases its affinity for the Arp2/3 complex and enhances actin polymerization by WASP. Mol Cell 2003; 11(5):1229-1239.
65. Suetsugu S, Hattori M, Miki H et al. Sustained activation of N-WASP through phosphorylation is essential for neurite extension. Dev Cell 2002; 3(5):645-658.
66. Cory GO, Garg R, Cramer R et al. Phosphorylation of tyrosine 291 enhances the ability of WASp to stimulate actin polymerization and filopodium formation. Wiskott-Aldrich Syndrome protein. J Biol Chem 2002; 277(47):45115-45121.
67. Torres E, Rosen MK. Contingent phosphorylation/dephosphorylation provides a mechanism of molecular memory in WASP. Mol Cell 2003; 11(5):1215-1227.
68. Eden S, Rohatgi R, Podtelejnikov AV et al. Mechanism of regulation of WAVE1-induced actin nucleation by Rac1 and Nck. Nature 2002; 418(6899):790-793.
69. Kitamura T, Kitamura Y, Yonezawa K et al. Molecular cloning of p125Nap1, a protein that associates with an SH3 domain of Nck. Biochem Biophys Res Commun 1996; 219(2):509-514.

70. Kobayashi K, Kuroda S, Fukata M et al. p140Sra-1 (specifically Rac1-associated protein) is a novel specific target for Rac1 small GTPase. J Biol Chem 1998; 273(1):291-295.
71. Dai Z, Pendergast AM. Abi-2, a novel SH3-containing protein interacts with the c-Abl tyrosine kinase and modulates c-Abl transforming activity. Genes Dev 1995; 9(21):2569-2582.
72. Stovold CF, Millard TH, Machesky LM. Inclusion of Scar/WAVE3 in a similar complex to Scar/WAVE1 and 2. BMC Cell Biol 2005; 6(1):11.
73. Basu D, Le J, El-Essal Sel D et al. DISTORTED3/SCAR2 is a putative arabidopsis WAVE complex subunit that activates the Arp2/3 complex and is required for epidermal morphogenesis. Plant Cell 2005; 17(2):502-524.
74. Kunda P, Craig G, Dominguez V et al. Abi, Sra1, and Kette control the stability and localization of SCAR/WAVE to regulate the formation of actin-based protrusions. Curr Biol 2003; 13(21):1867-1875.
75. Rogers SL, Wiedemann U, Stuurman N et al. Molecular requirements for actin-based lamella formation in Drosophila S2 cells. J Cell Biol 2003; 162(6):1079-1088.
76. Leng Y, Zhang J, Badour K et al. Abelson-interactor-1 promotes WAVE2 membrane translocation and Abelson-mediated tyrosine phosphorylation required for WAVE2 activation. Proc Natl Acad Sci USA 2005; 102(4):1098-1103.
77. Nakagawa H, Miki H, Nozumi M et al. IRSp53 is colocalised with WAVE2 at the tips of protruding lamellipodia and filopodia independently of Mena. J Cell Sci 2003; 116(Pt 12):2577-2583.
78. Soderling SH, Binns KL, Wayman GA et al. The WRP component of the WAVE-1 complex attenuates Rac-mediated signalling. Nat Cell Biol 2002; 4(12):970-975.
79. Deeks MJ, Hussey PJ. Arp2/3 and SCAR: Plants move to the fore. Nat Rev Mol Cell Biol 2005; 6(12):954-964.
80. Szymanski DB. Breaking the WAVE complex: The point of Arabidopsis trichomes. Curr Opin Plant Biol 2005; 8(1):103-112.
81. Blagg SL, Stewart M, Sambles C et al. PIR121 regulates pseudopod dynamics and SCAR activity in Dictyostelium. Curr Biol 2003; 13(17):1480-1487.
82. Blagg SL, Insall RH. Control of SCAR activity in Dictyostelium discoideum. Biochem Soc Trans 2004; 32(Pt 6):1113-1114.
83. Myers SA, Han JW, Lee Y et al. A Dictyostelium homologue of WASP is required for polarized F-actin assembly during chemotaxis. Mol Biol Cell 2005; 16(5):2191-2206.
84. Bogdan S, Klambt C. Kette regulates actin dynamics and genetically interacts with Wave and Wasp. Development 2003; 130(18):4427-4437.
85. Bogdan S, Stephan R, Lobke C et al. Abi activates Wasp to promote sensory organ. Nat Cell Biol 2005; 7(10):977-984.
86. Allen PG. Actin filament uncapping localizes to ruffling lamellae and rocketing vesicles. Nat Cell Biol 2003; 5(11):972-979.

Chapter 9

The Verprolins as Regulators of Actin Dynamics

Pontus Aspenström*

Abstract

Verprolin is an actin-binding protein first identified in budding yeast *Saccharomyces cerevisiae*. The yeast verprolin is needed for actin polymerisation during polarised growth and during endocytosis. In vertebrate cells, three genes encoding *Verprolin* orthologues have been identified: *WIP*, *CR16* and *WIRE/WICH*. The mammalian verprolins have been implicated in the regulation of actin dynamics either by binding directly to actin, by binding the WASP family of proteins or by binding to other actin regulating proteins. This review article will bring up to discussion the current understanding of the mechanisms underlying verprolin-dependent mobilisation of the actin filament system.

Introduction

Verprolin (Vrp1/end5) was found during a screen for a vinculin-like gene in budding yeast, *Saccharomyces cerevisiae*.[1] It turned out that, the very proline-rich protein (verprolin) identified this way had only limited similarity to vinculin, instead it has become the prototype for the verprolin family of proteins. Genes encoding verprolins have been identified in most eukaryotic organisms: fungi, nematodes and insects each have one gene-copy of *verprolin*, vertebrates have three genes encoding verprolin-like proteins.[2] In contrast, none of the plant genomes sequenced so far seem to encode a verprolin-like gene product. The verprolins have emerged as important effectors for signalling to actin dynamics mediated by the Wiskott-Aldrich syndrome protein (WASP) family of proteins. In addition, the verprolins can influence the actin polymerisation machinery in a manner independent of the WASP family of proteins. A general overview over the diverse functions of the verprolins was recently published.[2] Instead, this review article will focus on critical aspects of verprolin-dependent mobilisation of the actin filament system.

The Verprolins Have Multiple Binding Partners

The mammalian verprolins are known as WASP-interacting protein (WIP), glucocorticoid-regulated gene-product (CR16), and WIP-related (WIRE, also known as WIP and CR16 homologous protein, WICH).[3-6] All verprolins identified so far have a similar domain organisation (Fig. 1A). A consensus profilin-binding motif, XPPPPP in which X is a Gly, Ala, Leu, Ser or an Ile (also known as ABM-2 or PRM1 motif), is present in one or two copies in the N-terminus of all mammalian verprolins and WIP and WIRE have been shown to bind to profilin.[2,7,8] Furthermore, there are additional putative profilin-binding motifs in the mammalian verprolins: one in CR16, two in WIP and five in WIRE. The verprolins also contain

*Pontus Aspenström—Ludwig Institute for Cancer Research, Uppsala University, Biomedical Center, Box 595, SE-751 24 Uppsala, Sweden. Email: pontus.aspenstrom@LICR.uu.se

Actin-Monomer-Binding Proteins, edited by Pekka Lappalainen. ©2007 Landes Bioscience and Springer Science+Business Media.

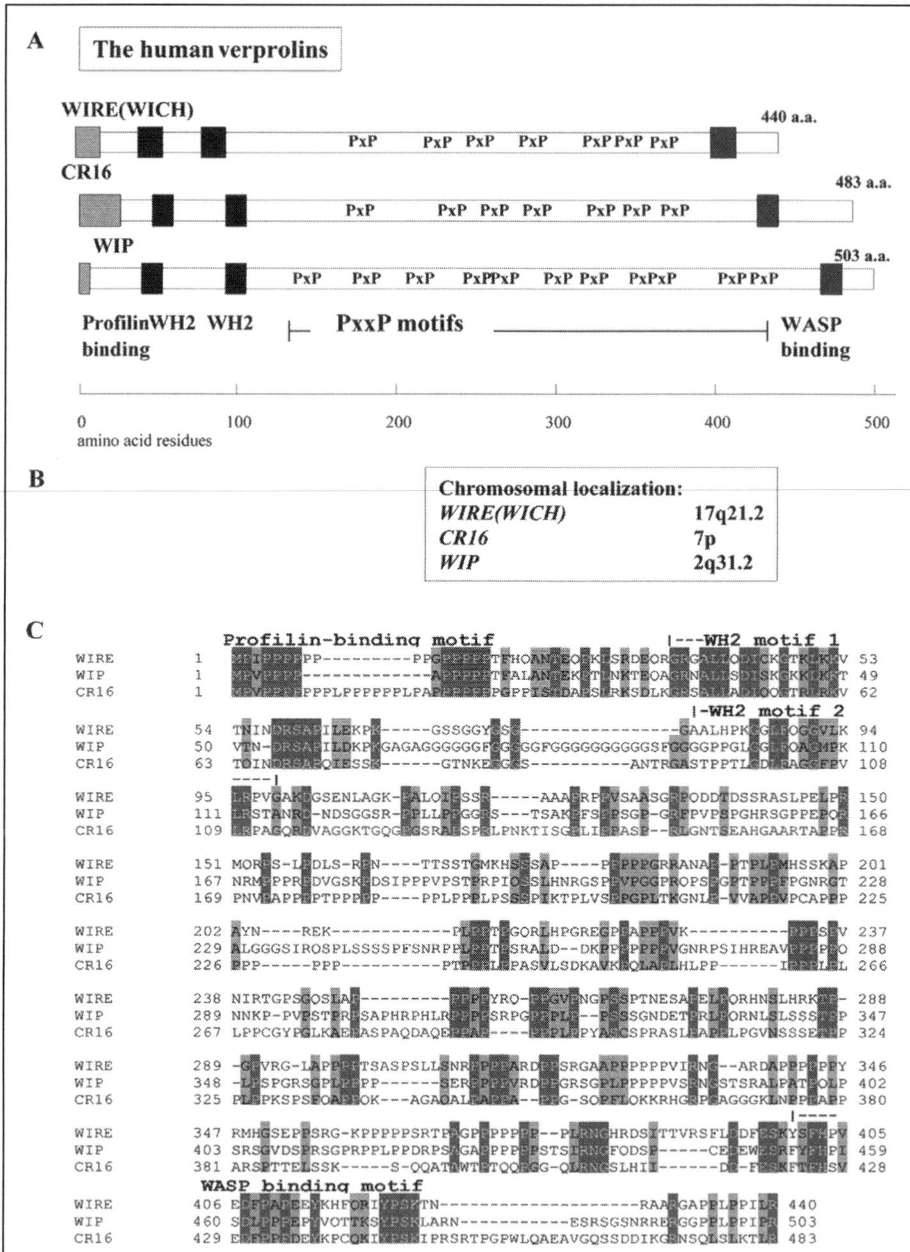

Figure 1. A) Schematic representation of the human verprolin family WIRE, CR16 and WIP. Light grey box = putative profilin binding motif, dark grey box = WASP binding motif, black box = WH2 motif. B) Chromosomal localisation of the human verprolin genes. C) Amino acid sequence alignment of all human verprolins. The positions of the potential profilin-binding motifs, the WH2 motifs and the WASP binding motif are marked in the sequence. The alignment was performed using the ClustalW algorithm.

N-terminal WASP homology 2 (WH2) motifs (Fig. 1A). This type of motif is found in a number of putative actin-binding proteins and it is supposed to confer binding to G-actin.[9]

The WASP-binding motif resides in the C-terminus of the verprolins and mediates the interaction to the WASP homology 1 (WH1) domain of the Wiskott-Aldrich Syndrome protein (WASP). Structural modelling has showed that the WH1 domains in WASP and in N-WASP are structurally related to the Ena/VASP homology 1 (EVH1) domain found in proteins, such as Mena, Ena and VASP.[10-12] EVH1 domains are known to bind a proline-rich consensus motif: (D/E)FPPPPX(D/E)(D/E), often referred to as an ABM-1 or PRM2 motif.[7,8] Similar sequence motifs found in the mammalian verprolins have been shown to be involved in WASP interaction (Fig. 1C).[14] In addition, aromatic amino acid residues N-terminally to this motif are also part of the WASP binding domain.[13,14]

All three members of the mammalian verprolins are extremely rich in prolines residues (proline content 27-29%), and in addition to binding profilin, the proline-rich motifs have been implicated in the interaction with SH3 domains and WW domains. The currently known binding partners have been collected into Table 1, however, it is likely that there exist additional binding partners for the verprolins that have not yet been identified.

Genetic Links to Actin Regulation

Most of the initial work on verprolin was performed employing budding yeast *S. cerevisiae* as a model organism. Budding yeast has turned out an ideal organism for the study of actin polymerisation in vivo and we owe a lot of our current insights into the regulation of actin dynamics to studies of budding yeast. In yeast cells, filamentous actin is organised in patches at the cell cortex and in cables.[15] The actin patches accumulate in the bud and at the site of cell-cell separation, as well as, on other sites of polarised cell surface growth, such as in the projecting so-called shmoos which are formed in yeast cells that are undergoing mating.[15] The actin patches appear to be formed from invaginations of the plasma membrane, covered with a coat of actin filaments.[16] Live cell imaging has demonstrated that the patches are highly dynamic and under a perpetual reconstruction during the phases of polarised growth.[17] Interestingly, disruption of the *VRP1* gene leads to a random distribution of actin patches, abolished chitin ring formation and decreased endocytosis of the α mating factor, suggesting that Vrp1p is essential for polarised growth of budding yeast.[18] All these phenotypes could, to a large extent, be explained by defects in actin polymerisation in the yeast cells lacking a functional *VRP1* gene.

Efforts made to identify the actin-binding domain in Vrp1p has resulted in conflicting observations. Vaduva et al showed that a fragment encompassing the N-terminal 70 amino-acid residues was sufficient for actin binding.[19] This fragment contain the first WH2 motif, and mutation analysis suggested a critical role for this motif in mediating the interaction to actin. In contrast, Thanabalu and Munn found that a fragment containing amino-acid residues 270-364 interacted with actin in the yeast-two hybrid assay.[20] These observations suggest the existence of several actin-binding motifs. Although WH2 motifs have been shown to bind globular actin (G-actin), the yeast two-hybrid assay is actually not likely to distinguish between G-actin and filamentous actin (F-actin).[9] Thus, the first WH2 motif might bind G-actin and additional motifs might bind F-actin. In line with this, Martinez-Quiles et al showed that a fragment of WIP encompassing amino-acid residues 1-127 bound to G-actin and, in addition, a full length WIP bound to F-actin in a cosedimentation assay.[21] It is possible that the artificial fragmentation of the molecule has separated motifs that are parts of the same interaction module in the native protein. Interestingly, the functional domains of Vrp1p has, to large extent, been conserved from yeast to mammals, since introduction of the mammalian WIP in yeast cells with the *VRP1* gene inactivated can suppress most of the phenotypes caused by *VRP1* inactivation.[22]

Several of the components involved in the Vrp1p-dependent actin polymerisation have been identified. Vrp1p has been shown to work in concert with the type I myosins Myo3p and Myo5p, the WASP homologue Las17/Bee1 and the Arp2/3 complex, to orchestrate the yeast actin organisation, as well as, endocytosis of the α mating factor receptor (also known

Table 1. Currently known verprolin binding partners

Verprolin Orthologue	Binding Partner	Cellular Function	Refs.
Saccharomyces cerevisiae			
Vrp1p (verprolin)	actin	cell migration, cell morphogenesis	19,20
	WASP orthologue		
	Las17p	actin organisation, endocytosis bud selection	24
	SH3 domain protein		
	Myo3p	actin organisation, endocytosis bud selection	23,25,26
	Myo5p	actin organisation, endocytosis bud selection	23,25,26
	Hof1p (Cyk2p)	cytokinesis, actomyosin ring formation	27,28
Caenorhabditis elegans			
	WASP orthologue		
WIP-1	WSP-1	cell migration, dorsal closure	44
Mammals			
WIP	G-actin, F-actin	cell migration, cell morphogenesis	21
	Profilin	actin polymerisation	3
	WASP orthologue		
	WASP	actin organisation	3
	N-WASP	actin organisation, endocytosis	21
	SH3 domain proteins		
	Cortactin	actin polymerisation	46
	Hck	nonreceptor tyrosine kinase	47
	Nck	adapter protein	45
	CrkL	T-cell receptor ligation	32
WIRE (WICH)	F-actin	cell migration, cell morphogenesis	30
	Profilin	actin polymerisation	5
	WASP orthologue		
	WASP	actin organisation	5,14
	N-WASP	actin organisation, endocytosis	5,14
	SH3 domain proteins		
	Nck	adapter protein	5
	Grb2	adapter protein	30
	WW domain proteins		
	CA150	?	48
	FE65	?	48
	WWOX	?	48
	WASP orthologue		
CR16	N-WASP	actin organisation, endocytosis	39

as Ste2p).[23-26] Vrp1p is also functioning during cytokinesis since growth of *vrp1* yeast is arrested at elevated temperatures. In addition, the protein is needed for the correct localisation of the FER CIP4 homology (FCH) domain-containing protein Hof1p/Cyk2p to the site of the bud.[27,28]

The Vertebrate Verprolins as Actin Binding Proteins

The actin binding capacity of the vertebrate verprolin has not been under any comprehensive study so far. Martinez-Quiles et al showed that WIP bound to G-actin as well as to filamentous actin.[21] Instead, most of the work on the verprolins in organisms bigger than yeast has been focused to the identification of functional modules of the verprolins, rather than studying actin binding per se. Ramesh et al in their initial report on WIP, showed that ectopic expression of WIP in a B-cell line resulted in an increase in the F-actin content in a WH2 motif-dependent manner.[3] Exogenously expressed WIRE, WIP and, to some extent, CR16 result in a increased bundling of actin filaments (Fig. 2).[14,29] In the case of WIRE, the appearance of F-actin bundles in WIRE-expressing cells was associated with a decreased amount of G-actin in the cells.[14] In agreement with this observation, WIRE (called WICH in this recent study) was shown to directly cross-link actin filaments, most likely via a fragment encompassing amino acid residues 167-381.[30] It is worth pointing out that the true nature of the actin bundles are not currently known. Are the actin bundles induced by the mammalian verprolins containing unipolar actin filaments or are they stress-fibres? Further studies are clearly necessary to give answers to these questions. Moreover, it is not know to what extent profilin contributes to the verprolin function.

Ectopically expressed WIP, WIRE and CR16 are able to shift the balance of platelet-derive growth factor (PDGF) stimulated porcine aortic endothelial cells stable transfected with the PDGF β-receptor (PAE/PDGFRβ) cells from the formation of edge ruffles to the formation of filopodia, indicating that the verprolins have roles in the formation of filopodia, rather than in the formation of lamellipodia (Fig. 2).[14] This effect was not mediated via WASP, as mutants of WIRE and WIP, which do not bind WASP, are still able to induce filopodia upon PDGF stimulation.[14] Furthermore, the effect was dependent on an intact WH2 motif, as a mutant

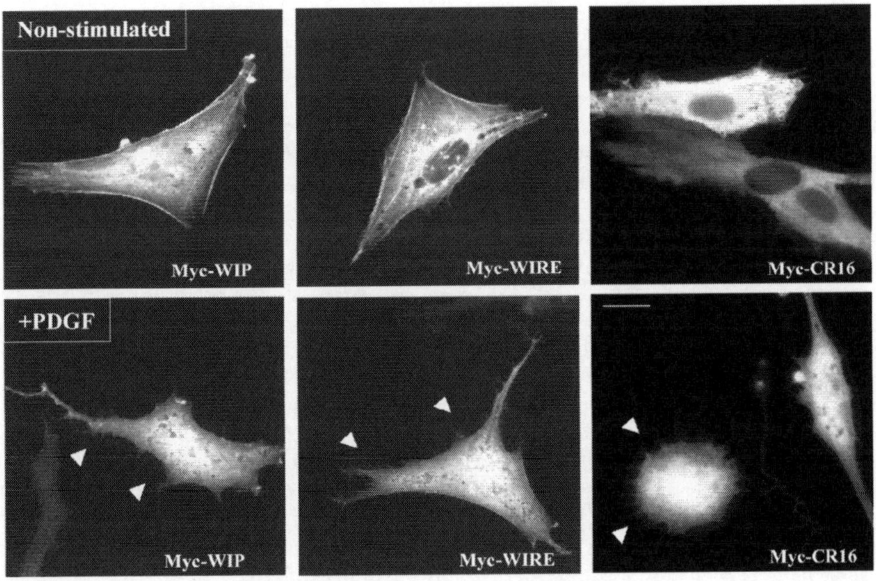

Figure 2. PAE/PDGFRβ cells transiently transfected with Myc-tagged human WIRE, human WIP and rat CR16.The localisation of the verprolins was detected by a mouse anti Myc-antibody followed by a FITC-conjugated anti mouse antibody. The cells in the lower panel were stimulated with 100 ng/ml PDGF-BB for 10 minutes in order to induce the formation of filopodia in the cells expressing the verprolins. The bar represents 20 μm.

WIRE with a dysfunctional first WH2 motif was unable to mediate PDGF-dependent filopodia formation.[14] One study on a fibroblast cell-line stably overexpressing exogenous WIP showed an increased formation of membrane ruffles, in particular dorsal ruffles, in response to PDGF.[31] Conversely, lung fibroblasts derived from WIP$^{-/-}$ mice have a much lower capacity to form dorsal ruffles compared to their wild-type counterparts. The enhanced ruffle formation was dependent on the actin-binding capacity of WIP, since a mutant WIP, in which the WH2 motifs had been deleted, was unable to mediate the PDGF-induced formation of dorsal ruffles, again pointing to the importance for the first WH2 motif in the regulation of actin dynamics.[31] All these observations implicate a role for the verprolins in the regulation of actin polymerisation, however it is currently not clear if the effect is dependent on the direct interaction between verprolins and actin or by interaction partners, such as the WASP family of proteins.

Verprolin and the WASP Family of Proteins

Many of the effects on actin polymerisation induced by the verprolins are mediated via the WASP family of proteins (WASP and N-WASP). In resting cells, 95% of N-WASP has been said to be sequestered by WIP, keeping N-WASP in an inactive conformation, (although this is only mentioned as an unpublished observation in this article by Sasahara et al).[32] It is not known if this is true also for WIRE or CR16. One study showed that Cdc42-dependent filipodia, triggered by microinjection of constitutively active Cdc42 or by treating the cells with bradykinin, required WIP since microinjection of an inhibitory anti-WIP antibody abrogated the filopodia formation.[21] WIP has also been found to be needed for the recruitment of N-WASP and the Arp2/3 during the intracellular motility of *Vaccinia* virus, leading to a model in which the tyrosine phosphorylated, virus-encoded, protein A36R, residing on the surface of the virus, recruits Nck, which, in turn, recruits WIP, N-WASP and the Arp2/3 complex, thereby inducing intracellular motility of the virus particles.[33,34] An Nck/WIP/N-WASP dependent pathway was also shown to be active during the formation of invadopodia in metastatic rat adenocarcinoma MTLn3 cells during invasive growth.[35] In yeast, the FCH domain-containing protein Hof1p/Cyk2p and Vrp1 were shown to collaborate during cytokinesis.[27] It is not known if the mammalian verprolins have a role during cytokinesis, but WIP was shown to bind the Cdc42-binding, FCH domain-containing protein Toca-1, which is related to the yeast protein Hof1p/Cyk2p.[36] This study showed that Toca-1 and the WIP:WASP complex work in concert during Cdc42-dependent actin assembly.

Verprolin and the Wiskott-Aldrich Syndrome

WASP is a protein that is found mutated in the rare X-linked immunodeficiency disorder Wiskott-Aldrich syndrome (WAS).[37,38] This disease is characterised by severe immunodeficiency affecting T- and B-cells, eczema and thrombocytopenia. In addition, patients have an increased incidence of lymphoreticular malignancies. WASP from patients harbours mutation in the WASP gene, and there is a prevalence of point mutations in the N-terminus of WASP and in particular in the WH1 domain of WASP.[37,38] These mutations have been shown to lead to a reduced or abolished binding to the verprolins, suggesting that they are critical mediators of the normal biological function of WASP.[39,40] WIP deficiency does not seem to lead to a syndrome with the severe immunodeficiency associated with the Wiskott-Aldrich syndrome, at least not in mice. WIP$^{-/-}$ mice are not dying prematurely and they display no clear differences from their wild-type littermates.[41] The lack of a strong phenotype in WIP$^{-/-}$ mice suggests that WIRE and CR16 can, compensate for the lack of WIP. However, no information on WIRE$^{-/-}$ or CR16$^{-/-}$ mice is currently available. Although WIP$^{-/-}$ mice have normal lymphocyte development, the WIP$^{-/-}$ T-cells have decreased abilities to proliferate, secrete IL-2, and change the cell shape in response to ligation of the T-cell receptor. T- and B-cells from WIP$^{-/-}$ mice have decreased amount of subcortical actin filaments.[41] In addition, WIP is needed for the signalling downstream of the high affinity receptor for IgE (FcϵRI) in masts cells, since it was shown that bone marrow-derived masts cells were impaired in their capacity to degranulate and secrete interleukin 6 after activation of the FcϵRI.[42] Moreover, the actin polymerisation occurring upon T-cell receptor ligation requires

WASP and is critical for T-cell activation.[32] T cell receptor ligation result in the formation of a ZAP-70/CrkL/WIP/WASP complex which is recruited to lipid rafts at the immunological synapses.[32] Mice with both the *WASP* and *WIP* genes inactivated are apparently viable, further indicating compensating functions of the CR16 and WIRE genes. However, chemotaxis to stromal cell-derived factor-1α (SDF-α) was severely impaired in the double knockout T-cells.[43] Have the mammalian verprolins essentially the same function but working in different cell-types? There is a difference in tissue distribution between the three members of the mammalian verprolins, however there are also indications that there are functional differences between them. For instance, WIRE is more effective than is WIP in blocking internalisation of the PDGFRβ.[14]

Concluding Remarks

We can now see a picture emerging over how the verprolins influence the actin polymerisation machinery, but there are still too many pieces missing to allow us to put the verprolins into a general context. There are several potential parallel pathways involving verprolins that lead to actin polymerisation and some of the possible routes are depicted in Figure 3. Activated tyrosine kinase receptors can potentially induce actin polymerisation via Nck, which binds to the

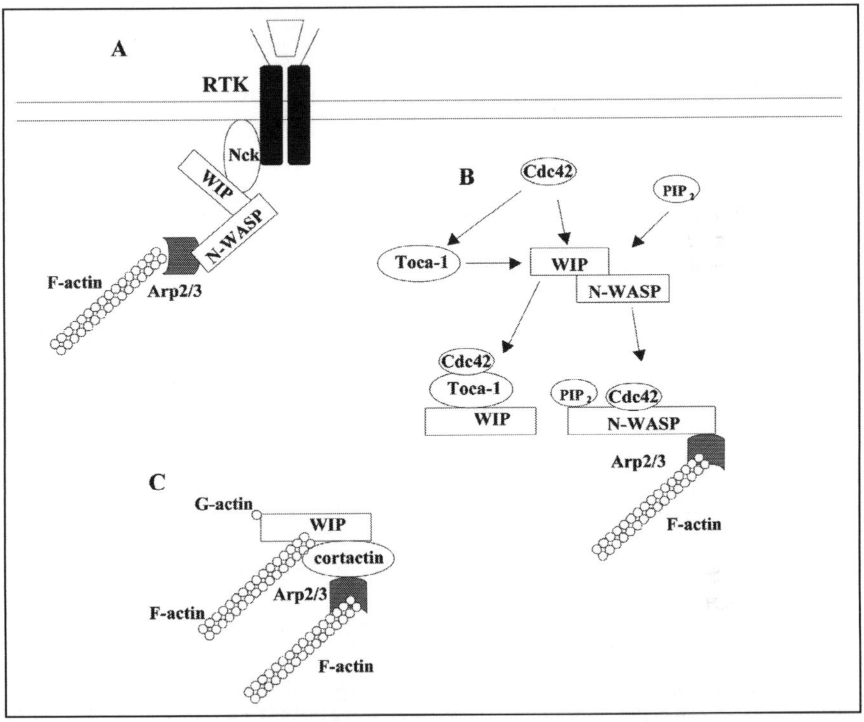

Figure 3. Schematic representation of the possible mechanisms for the WIP-regulated actin polymerisation in vertebrate cells. A) Activated tyrosine kinase receptors can induce actin polymerisation via an Nck- and WIP/N-WASP-dependent pathway to induce actin polymerisation. B) In resting cells, N-WASP is probably sequestered by WIP. Binding of Toca-1 to WIP can disrupt the WIP/N-WASP complex. The released N-WASP could then be activated by the concerted action of Cdc42 and phosphoinositides. The activated N-WASP can, in turn, induce actin polymerisation in an Arp2/3-dependent manner. C) WIP could induce actin polymerisation via an other binding partner, such as cortactin, which could induce Arp2/3-dependent actin polymerisation. In addition, WIP could affect the actin filament system by binding directly to either G-actin or F-actin.

activated receptor. Nck could then attract the WIP/N-WASP complex to induce actin polymerisation (Fig. 3A). In addition, WIP participate in N-WASP activation. In resting cells, N-WASP is thought to be sequestered in an active state by binding to verprolins such as WIP. Binding of Toca-1 to WIP might disrupt the WIP/N-WASP complex. The autoinhibited and released N-WASP could then be activated, possibly by binding to small GTPases, such as Cdc42, and to phosphoinositides to induce actin polymerisation (Fig. 3B). The released verprolins could, in turn, induce actin polymerisation via another binding partner, such as cortactin, which could induce Arp2/3 dependent actin polymerisation (Fig. 3C). WIP could also affect the actin filament system by direct binding to either G-actin or F-actin (Fig. 3C). Although, we do not see the exact mechanisms by which the verprolins communicate with the actin polymerisation machinery, it is obvious that they are essential players in the signalling pathways that link activated transmembrane receptors and the actin polymerisation machinery and, thereby, regulate the morphogenic and migratory abilities of vertebrate cells.

Acknowledgements

This work performed in the laboratory of the author has been supported in part by grants from the Swedish Cancer Society, Swedish Science Council. C.-H. Heldin is acknowledged for reading and commenting on the manuscript.

References

1. Donnelly SFH, Pocklington MJ, Pallotta D et al. A proline-rich protein, verprolin, involved in cytoskeletal organization and cellular growth in the yeast Saccharomyces cerevisiae. Mol Microbiol 1993;10:585-596.
2. Aspenström P. The verprolin family of proteins: Regulators of cell morphogenesis and endocytosis. FEBS Lett 2005; 579:5253-5259.
3. Ramesh N, Antón IM, Hartwig JH et al. WIP, a protein associated with Wiskott-Aldrich syndrome protein, induces actin polymerization and redistribution in lymphoid cells. Proc Natl Acad Sci USA 1997; 94:14671-14676.
4. Weiler MC, Smith JL, Masters JN. CR16, a novel proline-rich protein expressed in rat brain neurons, binds to SH3 domains and is a MAP kinase substrate. J Mol Neurosci 1996; 7:203-215.
5. Aspenström P. The WASP-binding protein WIRE has a role in the regulation of the actin filament system downstream of the platelet-derived growth factor receptor. Exp Cell Res 2002; 279:21-33.
6. Kato M, Miki H, Kurita S et al. WICH, a novel verprolin homology domain-containing protein that functions cooperatively with N-WASP in actin-microspike formation. Biochem Biophys Res Commun 2002; 291:41-47.
7. Holt MR, Koffer A. Cell motility: Proline-rich proteins promote protrusions. Trends Cell Biol 2001; 11:38-46.
8. Purich DL, Southwick FS. ABM-1 and ABM-2 homology sequences: Consensus docking sites for actin-based motility defined by oligoproline regions in Listeria ActA surface protein and human VASP. Biochem Biophys Res Commun 1997; 231:686-691.
9. Paunola E, Mattila PK, Lappalainen P. WH2 domain: A small, versatile adapter for actin monomers. FEBS Lett 2002; 513:92-97.
10. Callebaut I, Cossart P, Dehoux P. EVH1/WH1 domains of VASP and WASP proteins belong to a large family including Ran-binding domains of the RanBP1 family. FEBS Lett 1998; 441:181-185.
11. Prehoda KE, Lee DJ, Lim WA. Structure of the Enabled/Vasp homology 1 domain-peptide complex: A key component in the spatial control of actin assembly. Cell 1999; 97:471-480.
12. Volkman BF, Prehoda KE, Scott JA et al. Structure of the N-WASP EVH1 domain-WIP complex: Insight into the molecular basis of Wiskott-Aldrich Syndrome. Cell 2002; 111:565-576.
13. Zettl M, Way M. The WH1 and EVH1 domains of WASP and Ena/VASP family members bind distinct sequence motifs. Curr Biol 2002; 12:1-6.
14. Aspenström P. The mammalian verprolin homologue WIRE participates in receptor-mediated endocytosis and regulation of the actin filament system by distinct mechanisms. Exp Cell Res 2004; 298:485-498.
15. Schott D, Huffaker T, Bretscher A. Microfilaments and microtubules: The news from yeast. Curr Opin Microbiol 2002; 5:564-574.
16. Mulholland J, Preuss D, Moon A et al. Ultrastructure of the yeast actin cytoskeleton and its association with the plasma membrane. J Cell Biol 1994; 125:381-391.

17. Smith MG, Swamy SR, Pon LA. The life cycle of actin patches in mating yeast. J Cell Sci 2001; 114:1505-1513.
18. Munn AL, Stevenson BJ, Geli MI et al. end5, end6, and end7: Mutations that cause actin delocalization and block the internalization step of endocytosis in Saccharomyces cerevisiae. Mol Biol Cell 1995; 6:1721-1742.
19. Vaduva G, Martin NC, Hopper AK. Actin-binding verprolin is a polarity development protein required for the morphogenesis and function of the yeast actin cytoskeleton. J Cell Biol 1997; 139:1821-1833.
20. Thanabalu T, Munn AL. Functions of Vrp1p in citokinesis and actin patches are distinct and neither requires a WH2/V domain. EMBO J 2001; 24:6979-6989.
21. Martinez-Quiles N, Rohatgi R, Antón IM et al. WIP regulates N-WASP-mediated actin polymerization and filopodium formation. Nature Cell Biol 2001; 3:484-491.
22. Vaduva G, Martinez-Quiles N, Antón IM et al. The human WASP-interacting protein, WIP, activates the cell polarity pathway in yeast. J Biol Chem 1999; 274:17103-17108.
23. Anderson BL, Boldogh I, Evangelista M et al. The Src homology domain 3 (SH3) of a yeast type I myosin, Myo5p1, binds to verprolin and is required for targeting to sites of actin polarization. J Cell Biol 1998; 141:1357-1370.
24. Naqvi SN, Zahn R, Mitchell DA et al. The WASp homologue Las17p functions with the WIP homologue End5p/verprolin and is essential for endocytosis in yeast. Curr Biol 1998; 8:959-962.
25. Evangelista M, Klebl BM, Tong AHY et al. A role for myosin-I in actin assembly through interactions with Vrp1p, Bee1p, and the Arp2/3 complex. J Cell Biol 2000; 148:353-362.
26. Lechler T, Jonsdottir GA, Klee SK et al. A two-tiered mechanism by which Cdc42 controls the localization and activation of an Arp2/3-activating motor complex in yeast. J Cell Biol 2001; 155:261-270.
27. Naqvi SN, Feng Q, Boulton VJ et al. Vrp1p functions in both actomyosin ring-dependent and Hof1p-dependent pathways of cytokinesis. Traffic 2001; 2:189-201.
28. Ren G, Wang J, Brinkworth R et al. Verprolin cytokinesis function mediated by the Hof one trap domain. Traffic 2005; 6:575-593.
29. Vetterkind S, Miki H, Takenawa T et al. The rat homologue of Wiskott-Aldrich syndrome protein (WASP)-interacting protein (WIP) associates with actin filaments, recruits N-WASP from the nucleus, and mediates mobilization of actin from stress fibers in favor of filopodia formation. J Biol Chem 2002; 277:87-95.
30. Kato M, Takenawa T. WICH, a member of WASP-interacting protein family, cross-links actin filaments. Biochem Biophys Res Commun 2005; 328:1058-1066.
31. Antón IM, Saville SP, Byrne MJ et al. WIP participates in actin reorganization and ruffle formation induced by PDGF. J Cell Sci 2003; 116:2443-2451.
32. Sasahara Y, Rachid R, Byrne MJ et al. Mechanism of recruitment of WASP to the immunological synapse and of its activation following TCR ligation. Mol Cell 2002; 10:1269-1281.
33. Frischknecht F, Moreau V, Röttger S et al. Actin-based motility of vaccinia virus mimics receptor tyrosine kinase signalling. Nature 1999; 401:926-929.
34. Moreau V, Frischknecht F, Reckmann I et al. A complex of N-WASP and WIP integrates signalling cascades that lead to actin polymerization. Nature Cell Biol 2000; 2:441-448.
35. Yamaguchi H, Lorenz M, Kempiak S et al. Molecular mechanisms of invadopodium formation: The role of the N-WASP-Arp2/3 complex pathway and cofilin. J Cell Biol 2005; 168:441-452.
36. Ho HY, Rohatgi R, Lebensohn AM et al. Toca-1 mediates Cdc42-dependent actin nucleation by activating the N-WASP-WIP complex. Cell 2004; 118:203-216.
37. Badour K, Zhang J, Siminovitch KA. The Wiskott-Aldrich syndrome protein: Forging the link between actin and cell activation. Immunol Rev 2003; 192:98-112.
38. Burns S, Cory GO, Vainchenker W et al. Mechanisms of WASp-mediated hematologic and immunologic disease. Blood 2004; 104:3454-3462.
39. Stewart DM, Tian L, Nelson DL. Mutations that cause the Wiskott-Aldrich syndrome impair the interaction of Wiskott-Aldrich syndrome protein (WASP) with WASP interacting protein. J Immunol 1999; 162:5019-5024.
40. Ho HYH, Rohatgi R, Ma L et al. CR16 forms a complex with N-WASP in brain and is a novel member of a conserved proline-rich actin-binding protein family. Proc Natl Acad Sci USA 2001; 98:11306-11311.
41. Antón IM, de la Fuente MA, Sims TN et al. WIP deficiency reveals a differential role for WIP and the actin cytoskeleton in T and B cell activation. Immunity 2002; 16:193-204.
42. Kettner A, Kumar L, Antón IM et al. WIP regulates signalling via the high affinity receptor for immunoglobulin E in masts cells. J Exp Med 2004; 199:357-368.

43. Gallego MD, de la Fuente MA, Antón IM et al. WIP and WASP play complementary roles in T cell homing and chemotaxis to SDF-1α. Int Immunol 2005.
44. Sawa M, Takenawa T. Caenorhabditis elegans WASP-interacting protein homologue WIP-1 is involved in morphogenesis through maintenance of WSP-1 protein levels. Biochem Biophys Res Commun 2006; 340:709-717.
45. Antón IM, Lu W, Mayer BJ et al. The Wiskott-Aldrich syndrome protein-interacting protein (WIP) binds to the adaptor protein Nck. J Biol Chem 1998; 273:20992-20995.
46. Kinley AW, Weed SA, Weaver AM et al. Cortactin interacts with WIP in regulating Arp2/3 activation and membrane protrusion. Curr Biol 2003; 13:384-393.
47. Scott MP, Zappacosta F, Kim EY et al. Identification of novel SH3 domain ligands for the Src family kinase Hck. Wiskott-Aldrich syndrome protein (WASP), WASP-interacting protein (WIP), and ELMO1. J Biol Chem 2002; 277:28238-28246.
48. Ingham RJ, Colwill K, Howard C et al. WW domains provide a platform for the assembly of multiprotein complexes. Mol Cell Biol 2005; 25:7092-7106.

Chapter 10

A Common Binding Site for Actin-Binding Proteins on the Actin Surface

Roberto Dominguez*

Abstract

The dynamic remodeling of the actin cytoskeleton plays an essential role in many cellular processes, including cell motility, cytokinesis, and intracellular transport. A large number of actin-binding proteins (ABPs) participate in this process, regulating the assembly of actin filaments into functional networks. ABPs are extremely diverse, both structurally and functionally, but they most seem to share a common binding area on the actin surface, consistent of the cleft between actin subdomains 1 and 3. Actin itself is thought to interact in this cleft in the filament. As a result, part of the cleft becomes buried in F-actin by inter-subunit contacts, whereas another part remains exposed and mediates the interactions of various filamentous actin-binding proteins. The convergence of actin-binding proteins into a common binding area imposes enormous constraints on their interactions and could serve a regulatory function. Because the cleft falls near the hinge for domain motions in actin, binding in this area is an effective way for ABPs to "sense" the conformation of actin, in particular conformational changes resulting from ATP hydrolysis by actin or from the G- to F-actin transition.

Introduction

The dynamic remodeling of the actin cytoskeleton is essential for many cellular functions, including motility, cytokinesis, and the control of cell shape and polarity.[1] Actin is the major component of the cytoskeleton. It exists in two different forms, a monomeric form (G-actin) and a filamentous form (F-actin). F-actin is structurally and functionally asymmetric, undergoing net association of ATP-actin monomers to the barbed end (+ end) and dissociation of ADP-actin monomers from the pointed end (- end), a process known as actin filament treadmilling. In vivo, the transition between G- and F-actin is tightly regulated by a vast number of actin-binding proteins (ABPs). These proteins direct the location, rate, and timing for actin assembly into different cytoskeletal networks such as filopodia, lamellipodia and focal adhesions.[2-4] Typically, ABPs are multidomain proteins containing, in addition to their actin-binding domains, signaling domains and protein-protein interaction modules. Although the number of ABPs is extremely large and is constantly growing, the actin-binding domains of most ABPs can be grouped into structurally conserved folding families, including the WASP homology domain-2 (WH2),[5] the actin-depolymerizing factor/cofilin (ADF/cofilin) domain,[6] the gelsolin-homology domain,[7] the calponin-homology (CH) domain,[8] the formin homology 2 (FH2) domain[9] and the myosin motor domain,[10] to mention just a few. The structures of complexes of actin with some members of these folding families are now known, and are

*Roberto Dominguez—Boston Biomedical Research Institute, 64 Grove Street, Watertown, Massachusetts 02472, U.S.A. Email: RDominguez@bbri.org

Actin-Monomer-Binding Proteins, edited by Pekka Lappalainen. ©2007 Landes Bioscience and Springer Science+Business Media.

starting to reveal common features in the way ABPs interact with actin.[11] Such common structural features and their functional implications are discussed here.

A Prevalent Target-Binding Cleft in Actin

From yeast to human, actin is one of the most highly conserved proteins in nature.[12] It consists of 375 amino acids. The molecule is organized into two structurally related domains, which are thought to have resulted from gene duplication.[13] The two domains can be further subdivided into subdomains 1 to 4 (Fig. 1A). Two diametrically opposed clefts separate the two large domains of actin. The larger cleft, between subdomains 2 and 4, constitutes the nucleotide-binding site, whereas the smaller cleft, between subdomains 1 and 3, mediates the interactions of actin with most ABPs (Fig. 1B).[11] Thus, all the structures of complexes of actin with ABPs, except that of actin-DNase I,[13] interact in this cleft (Fig. 2). Profilin also interacts

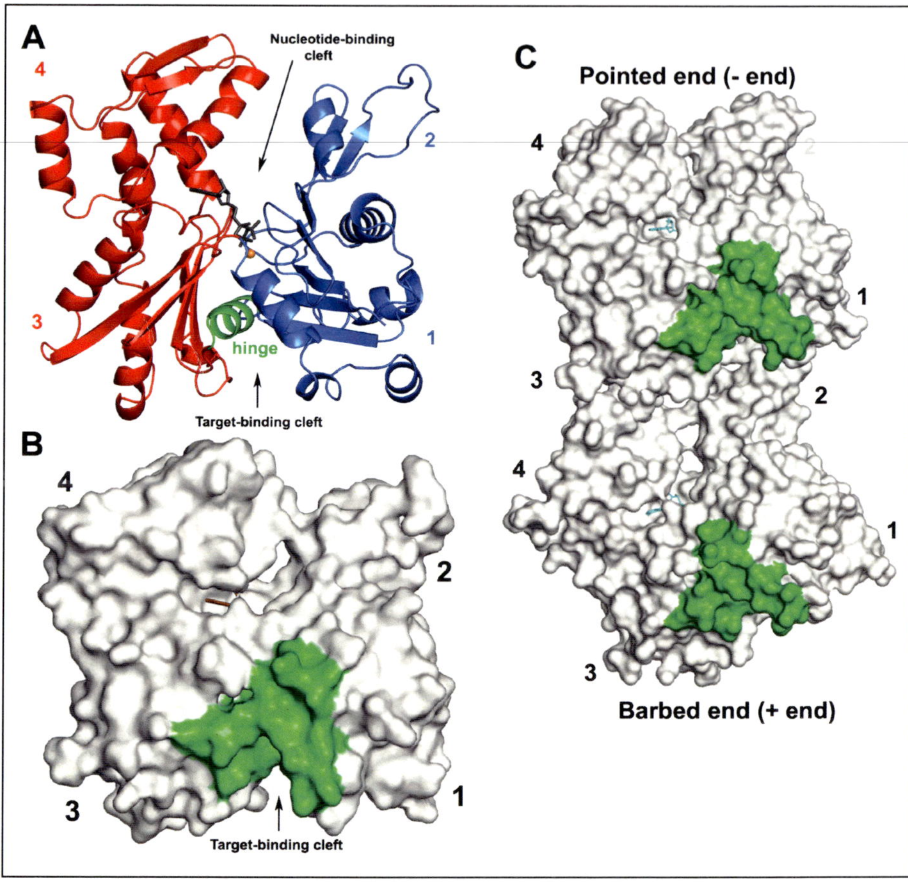

Figure 1. A prevalent target-binding cleft in actin. A) Ribbon representation illustrating the "conventional" view of actin. Two diametrically opposed clefts, the nucleotide cleft and the target-binding cleft, effectively separate the actin molecule into two large domains. The polypeptide chain goes across domains only twice. The α-helix between amino acids Ile 136 and Gly 146, shown in green, serves as a hinge for inter-domain motions. This α-helix lines the target-binding cleft. B) Surface representation of actin (same orientation as in part A), showing the target-binding cleft in green. C) Two molecules of actin along one filament strand of Holmes' F-actin model.[43]

to the back of this cleft in actin,[14] although in a different way than most ABPs. The specifics of these interactions are discussed below.

Gelsolin

Gelsolin is a calcium regulated F-actin capping and severing protein.[7] In the structure of the complex of gelsolin fragment 1 with actin,[15,16] the major contact involves α-helix Ser 70-Asn 89 of gelsolin, which binds in the cleft between subdomains 1 and 3 of actin (Fig. 2). Actin residues Tyr 143, Ala 144, Gly 146, Thr 148, Gly 168, Ile 341, Ile 345, Leu 346, Leu 349, Thr 351, Met 355, and possibly the C-terminus of actin, which is typically disordered in the structures, line the cleft in actin. The interaction is mostly hydrophobic, although a few hydrogen bonds are also observed. The α-helix in gelsolin presents exposed hydrophobic side chains (Fig. 2), which interact with the hydrophobic amino acids that line the cleft in actin.

Vitamin D-Binding Protein (DBP)

A similar interaction was later observed in the structure of the complex of actin with DBP.[17] Similar to gelsolin, DBP α-helix Ser194-Asp204 binds in the cleft in actin (Fig. 2). Although DBP interacts with actin over a large interface, the interaction involving this α-helix appears to play a predominant role in the formation of their complex. The presence of a common actin-binding motif in these two otherwise unrelated proteins was proposed to allow DBP to compete effectively for actin monomer binding, while freeing gelsolin for its severing function as part of the actin-scavenger system.[17]

Thymosin-β Domain

Thymosin-β4 is the prototypical member of the thymosin β family.[4,18] Tβ4 is a small (5-kDa) actin monomer sequestering protein, which constitutes an important buffer of ATP-actin monomers in the cell. The actin-bound structures of a hybrid protein, consisting of gelsolin domain 1 and the C-terminal half of Tβ4[19] and that of the N-terminal half of ciboulot domain 1,[20] a Tβ4-related molecule from *Drosophila megalonaster*, have been determined. The N-terminal portion of ciboulot and the C-terminal portion of Tβ4 from these two structures connect rather well to produce a model of the complex of Tβ-actin[19] (Fig. 2). Combined, these structures reveal that the Tβ domain consists of two α-helices, an N- and a C-terminal α-helices, connected by a linker. The N- and a C-terminal α-helices cap the barbed and the pointed end of actin, respectively, providing a structural explanation for the monomer sequestering function of the Tβ domain. The N-terminal α-helix binds in the cleft between actin subdomains 1 and 3 (Fig. 2). The binding of the α-helix in ciboulot, however, differs from that of gelsolin and DBP in one important way. While the α-helix in ciboulot runs from back to front (according to the conventional view in (Fig. 1A), those in gelsolin and DBP run from front to back. A superimposition of the structures, using actin as a reference, reveals that despite the different directionalities of binding, some of the hydrophobic side chains in the α-helices of these three proteins occupy similar positions within the cleft in actin. Note, however, that there is no significant sequence similarity between the α-helices of these proteins. The only common feature is the periodicity of exposed hydrophobic amino acids on one side of the α-helix.

WASP Homology 2 (WH2) Domain

It had been proposed, based on sequence analysis, that the WH2 domain and the Tβ domain formed part of an extended family.[5] However, this view remained controversial, in part because of the different biological functions and low sequence similarity between the two domains.[21] Moreover, Tβ proteins consist of one or multiple copies of the Tβ domain, whereas the WH2 domain is found within large multidomain proteins, typically in the form of tandem repeats. The recent determination of the structures of actin complexes with the WH2 domains of various cytoskeletal proteins, including the prototypical WH2 of WASP, confirmed to a large extent the proposed relationship.[22] In particular, the WH2 domain presents and N-terminal

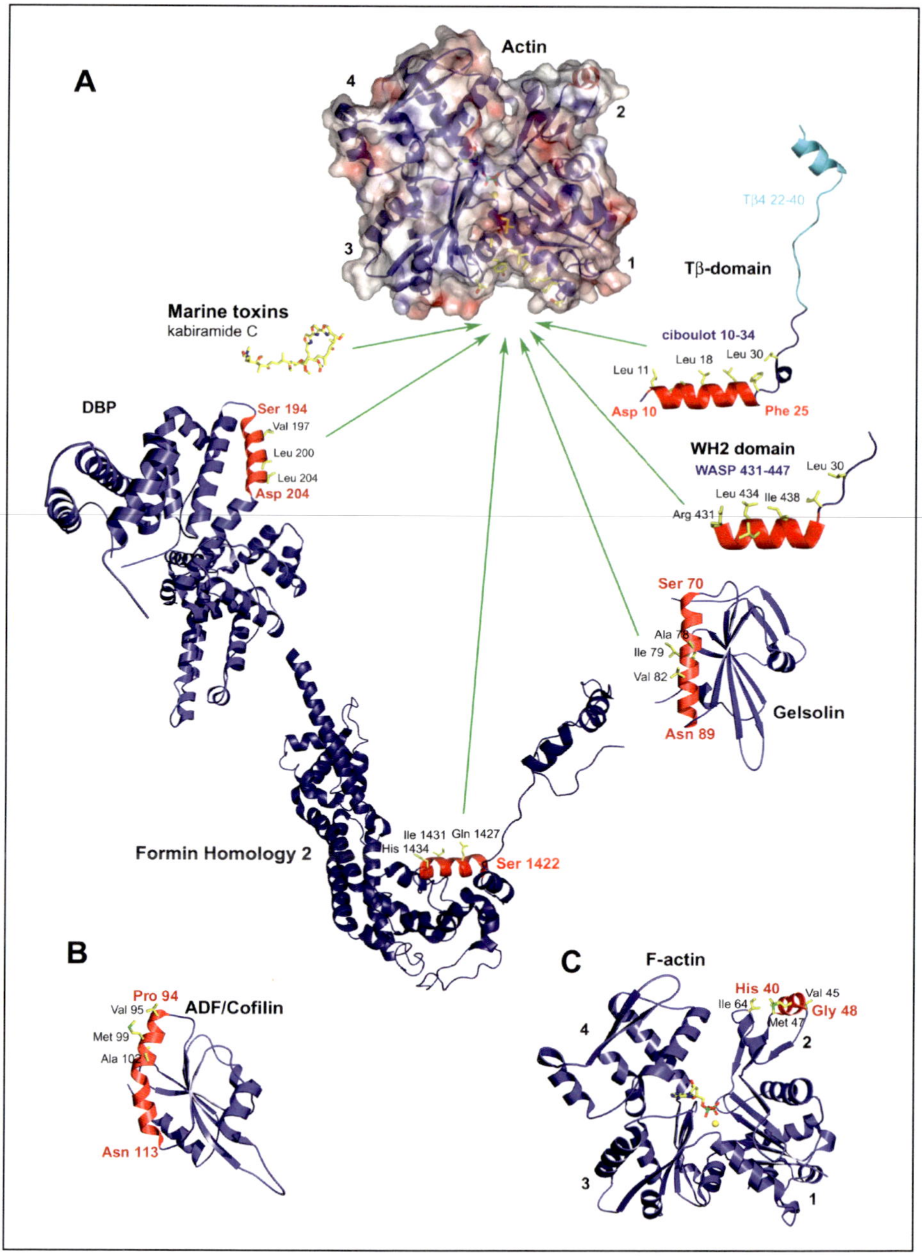

Figure 2. The cleft between actin subdomains 1 and 3 is a hot spot for actin-binding proteins. A) Illustration of all the ABPs known to present and α-helix (shown in red) that interacts in the target-binding cleft in actin. An electrostatic surface representation of actin is shown on top of the ribbon diagram to illustrate the global hydrophobicity of the cleft. The structures shown are (counterclockwise): DBP,[44] formin,[24] gelsolin,[15] WASP WH2 domain,[22] and the Tβ domain.[19,20] The hydrophobic cleft is also targeted by marine toxins, such as kabiramide C and jaspisamide A.[26] B,C) ADF/cofilin and actin are also predicted to interact in this cleft.

α-helix whose hydrophobic side interacts in the cleft between actin subdomains 1 and 3. Here again, although there is no significant sequence conservation between the α-helices of Tβ and WH2, the periodicity of hydrophobic amino acids involved in the interaction with actin is well conserved. However, there are also important differences between the Tβ and WH2 domains. Notably, the WH2 domain lacks the C-terminal α-helix characteristic of the Tβ domain, which is consistent with the role of tandem WH2s in actin filament nucleation.[22,23]

Formin Homology 2 (FH2) Domain

Formins are a family of modular proteins that mediate the nucleation and elongation of unbranched actin filaments.[9] The FH2 domain binds actin filament barbed ends and moves processively as the filaments elongate or depolymerize. The crystal structure of the FH2 domain of Bni1p in complex with actin has been determined recently.[24] In the structure, FH2 forms a dimer, with each subunit interacting with two actin molecules in an orientation similar to that of the double-stranded barbed-end of the filament, consistent with the formation of a filament nucleus. Formin also turned out to have an α-helix that binds in the cleft between actin subdomains 1 and 3 (amino acids Ser 1422 to His 1434) (Fig. 2). The orientation of the α-helix in the complex with formin is similar to that of the WH2 and Tβ domains. Interestingly, the α-helix in formin only exposes a single hydrophobic amino acid (Ile 1431), with the rest of the interaction having a polar character.

Toxofilin

Toxofilin is an actin sequestering protein from *Toxoplasma gondii.*[25] This parasite displays a strikingly low amount of actin filaments, suggesting that actin monomer sequestration may play a key role in parasite actin dynamics. The structure of the complex toxofilin-actin is being determined in our lab. Strikingly, toxofilin also presents an α-helix, similar to that of the WH2 domain, that binds in the cleft in actin (Lee et al, in preparation).

Marine Toxins

Actin also binds a series of drugs and toxins, including cytochalasins, phallotoxins, macrrolide toxins and marine macrolide toxins.[12] The structures of actin complexed with the marine toxins kabiramide C and jaspisamide A reveal that the binding site for these molecules is also the cleft between actin subdomains 1 and 3.[26]

Conformational Plasticity of the Target-Binding Cleft in Actin

The portion of the α-helix in ciboulot that interacts with actin is longer than in gelsolin and DBP. Ciboulot has exposed hydrophobic side chains along four consecutive helical turns (Fig. 2). These side chains bind in the hydrophobic cleft in actin, covering the entire length of the cleft. The N-terminus of the α-helix in ciboulot partially overlaps with the binding site of profilin,[14] which binds to the back of the cleft in the standard view, although also interacting with subdomains 1 and 3. In contrast, Tβ4, whose α-helix is predicted to be shorter than that of ciboulot,[27] can bind actin simultaneously with profilin.[28] In gelsolin, DBP and formin, the portion of the α-helix that interacts with actin is even shorter, and binds only to the front half of the cleft. The orientation of the α-helices of gelsolin and DBP in the cleft is also opposite to that of formin, Tβ and WH2. Thus, the hydrophobic cleft in actin is long enough to accommodate interactions at its front or back halves, or throughout its entire length. This opens the possibility that two ABPs, whose binding sites on the cleft do not fully overlap, could bind to actin simultaneously, either transiently or as a stable complex. This appears to be the case for Tβ4 and profilin,[28] as well as for certain WH2 domains and profilin (Chereau and Dominguez, J Struct Biol, in press). A similar situation can take place in the filament. Indeed, it is possible that in F-actin the D-loop of an actin subunit binds to the back of the hydrophobic cleft of a neighboring subunit, an interaction that can be in addition regulated by nucleotide hydrolysis by actin (Fig. 1C). This would explain how tandem WH2 domains can coexist with and nucleate actin filaments, as observed in spire.[23] ABPs that bind at the front end of the cleft could either coexist with F-actin, compete with F-actin, or take advantage of nucleotide-dependent conformational changes in actin to access the cleft, possibly severing the filament. Coincidently, both Tβ4[29] and ADF/cofilin,[30] which as proposed below may also bind in the cleft, change the twist of F-actin upon binding, suggesting competition with some of the inter-subunit contacts in F-actin. It appears therefore that the hydrophobic cleft in actin is highly adaptable, it can accommodate interactions with a range of unrelated ABPs. These interactions typically involve an α-helix in the ABP, but the general position and orientation of the α-helix varies. It also appears plausible that certain ABP could bind to different parts of the cleft simultaneously.

Crosstalk between the Target-Binding and Nucleotide Clefts

As explained above, the nucleotide cleft and the target-binding cleft effectively divide actin into two large domains (Fig. 1A). The polypeptide chain goes across domains only twice, and the point of intercept between the two domains constitutes the hinge for domain motions in actin, physically coinciding with the α-helix Ile136 to Gly146[31,32] (Fig. 1A). In this way, nucleotide-dependent conformational changes in actin can be sensed by ABPs, explaining the strong correlation existing between the state of the nucleotide and the actin-binding affinities of most ABPs. The crosstalk between clefts is most likely also responsible for the inhibition of nucleotide hydrolysis resulting from the binding of many ABPs. Interestingly, a similar two-cleft system exists in myosin, where a hinge separates the nucleotide-binding cleft from the actin-binding cleft. In this way, nucleotide-dependent movements are physically transmitted to the actin-binding site, thereby modulating the actin-binding affinity of myosin.[33]

Implication for Other Actin-Binding Proteins

The evidence to date is consistent with the hydrophobic cleft in actin being a primary target for ABPs. Other proteins, whose binding sites on actin remain unknown, may also bind in this cleft. Any protein that binds in this cleft will most likely contain an α-helix, featuring few exposed and conserved hydrophobic amino acids, equivalent to those of gelsolin, DBP, WH2 and Tβ domains.

ADF/Cofilin

Such a conserved α-helix exists among members of the ADF/cofilin family.[6] Although structures of various members of this family, including destrin,[34] cofilin,[35] actophorin,[36] ADF1,[37] and the N-terminal domain of twinfilin[38] have been reported, a structure of a complex with actin has remained elusive. There is ample evidence linking ADF/cofilin helix 3 with actin binding.[38-41] There is also a theoretical model of the actin-cofilin complex that takes into account the existing general structural similarity between cofilin and gelsolin and proposes a similar mode of binding for these two proteins.[42] Because of the importance of ADF/cofilin helix 3 in actin binding,[38-41] the fact that it is one of the most highly conserved regions in this family, and the presence of exposed hydrophobic side chains (Fig. 2B), it is plausible that ADF/cofilin α-helix 3 also binds in the hydrophobic cleft in actin (Fig. 2B). Due to the overall similarity between the ADF/cofilin[37] and gelsolin[7] folds, it is further possible that the directionality of binding of ADF/cofilin α-helix 3 is the same as in gelsolin and DBP, i.e., from front to back.

F-Actin

The hydrophobic cleft also appears to be involved in inter-subunit contacts in F-actin,[43] raising the exciting possibility that ABPs compete with actin for this binding site. This may require the presence of an α-helix, containing exposed hydrophobic side chains within actin itself. Moreover, changes to this α-helix or the hydrophobic cleft would be expected to affect actin assembly. Consistent with this idea, proteins that are known to bind in the cleft often block actin polymerization. Moreover, actin labeled at Cys 374 with tetramethylrhodamine-5-maleimide (TMR) becomes polymerization deficient, allowing TMR-actin to be crystallized in the absence of any bound protein.[44] The resulting structure reveals the TMR probe partially blocking the cleft in actin, which could explain the dramatic effect that this probe has in polymerization. The most likely candidate to bind in the hydrophobic cleft in F-actin is the DNase I-binding loop (D-loop, amino acids His 40-Gly 48) of a neighboring actin subunit (Fig. 2C). As its name indicates, the D-loop mediates the formation of the strong actin-DNase I complex.[13] Labeling[45] or cleavage[46,47] of the D-loop affects actin polymerization. Furthermore, the D-loop can be directly cross-linked to Cys 374 in the cleft of an adjacent monomer within an F-actin strand.[48,49] Therefore, the existing evidence suggests that in F-actin the D-loop of an actin monomer binds in the hydrophobic cleft of a neighboring monomer. However, in most actin structures the D-loop appears either disordered or folded as an extended β-hairpin loop, not an α-helix. Interestingly, in one of the structures, that of TMR-actin in the ADP state, the D-loop adopts an α-helical conformation.[44] The α-helix in the D-loop, which had not been observed before, was assumed to be part of the global nucleotide-dependent conformational change.[32,44] However, it is also possible that this conformation is only stable if the D-loop is in contact with a binding partner (or a neighboring molecule in the crystal as it is the case in this structure). Thus, this structure may have provided the first glance of the structure of the D-loop in the filament, where its conformation may be constrained by inter-subunit interactions, which could favor the α-helical conformation. Independent of these considerations, an updated model of the actin filaments has been proposed that positions the loop, in the α-helical conformation observed in the ADP-TMR-actin structure, in close proximity of the cleft of the next subunit in the filament strand.[50] The presence of the D-loop at this location would result in steric hindrance with TMR bound at Cys 374, possibly explaining the negative effect that this probe has on polymerization.

EM reconstructions of F-actin decorated with various actin-binding proteins, including myosin,[50,51] cofilin,[30] and the ABD domains of various members of the spectrin family,[52-56] all show density masking the cleft in F-actin.[57] It is therefore likely that these proteins all present specific interactions with the cleft in actin. Because actin, and the hydrophobic pocket in particular, are highly conserved from yeast to human, a potentially powerful way to determine the corresponding binding interface of F-actin-binding partners is to plot sequence conservation on the surface of high-resolution structures. Proteins typically tolerate significant sequence variation on their surface, but the F-actin binding function would be expected to force sequence conservation at the binding interface.

Acknowledgements

Supported by NIH grant GM073791. The author thanks Francois Ferron for help with the preparation of Figure 1.

References

1. Pollard TD, Borisy GG. Cellular motility driven by assembly and disassembly of actin filaments. Cell 2003; 112(4):453-465.
2. Paavilainen VO, Bertling E, Falck S et al. Regulation of cytoskeletal dynamics by actin-monomer-binding proteins. Trends Cell Biol 2004; 14(7):386-394.
3. Lambrechts A, Van Troys M, Ampe C. The actin cytoskeleton in normal and pathological cell motility. Int J Biochem Cell Biol 2004; 36(10):1890-1909.
4. dos Remedios CG, Chhabra D, Kekic M et al. Actin binding proteins: Regulation of cytoskeletal microfilaments. Physiol Rev 2003; 83(2):433-473.
5. Paunola E, Mattila PK, Lappalainen P. WH2 domain: A small, versatile adapter for actin monomers. FEBS Lett 2002; 513(1):92-97.

6. Lappalainen P, Kessels MM, Cope MJ et al. The ADF homology (ADF-H) domain: A highly exploited actin-binding module. Mol Biol Cell 1998; 9(8):1951-1959.
7. McGough AM, Staiger CJ, Min JK et al. The gelsolin family of actin regulatory proteins: Modular structures, versatile functions. FEBS Lett 2003; 552(2-3):75-81.
8. Gimona M, Djinovic-Carugo K, Kranewitter WJ et al. Functional plasticity of CH domains. FEBS Lett 2002; 513(1):98-106.
9. Higgs HN. Formin proteins: A domain-based approach. Trends Biochem Sci 2005; 30(6):342-353.
10. Sellers JR. Myosins: A diverse superfamily. Biochim Biophys Acta 2000; 1496(1):3-22.
11. Dominguez R. Actin-binding proteins—A unifying hypothesis. Trends Biochem Sci 2004; 29(11):572-578.
12. Sheterline P, Clayton J, Sparrow J. Actin. 4th ed. Protein Profile 1998; 2(1):1-272.
13. Kabsch W, Mannherz HG, Suck D et al. Atomic structure of the actin: DNase I complex. Nature 1990; 347(6288):37-44.
14. Schutt CE, Myslik JC, Rozycki MD et al. The structure of crystalline profilin-beta-actin. Nature 1993; 365(6449):810-816.
15. McLaughlin PJ, Gooch JT, Mannherz HG et al. Structure of gelsolin segment 1-actin complex and the mechanism of filament severing. Nature 1993; 364(6439):685-692.
16. Irobi E, Burtnick LD, Urosev D et al. From the first to the second domain of gelsolin: A common path on the surface of actin? FEBS Lett 2003; 552(2-3):86-90.
17. Otterbein LR, Cosio C, Graceffa P et al. Crystal structures of the vitamin D-binding protein and its complex with actin: Structural basis of the actin-scavenger system. Proc Natl Acad Sci USA 2002; 99(12):8003-8008.
18. Bubb MR. Thymosin beta 4 interactions. Vitam Horm 2003; 66:297-316.
19. Irobi E, Aguda AH, Larsson M et al. Structural basis of actin sequestration by thymosin-beta4: Implications for WH2 proteins. EMBO J 2004; 23(18):3599-3608.
20. Hertzog M, van Heijenoort C, Didry D et al. The beta-thymosin/WH2 domain; structural basis for the switch from inhibition to promotion of actin assembly. Cell 2004; 117(5):611-623.
21. Edwards J. Are beta-thymosins WH2 domains? FEBS Lett 2004; 573(1-3):231-232, (author reply 233).
22. Chereau D, Kerff F, Graceffa P et al. Actin-bound structures of Wiskott-Aldrich syndrome protein (WASP)-homology domain 2 and the implications for filament assembly. Proc Natl Acad Sci USA 2005; 102(46):16644-16649.
23. Quinlan ME, Heuser JE, Kerkhoff E et al. Drosophila Spire is an actin nucleation factor. Nature 2005; 433(7024):382-388.
24. Otomo T, Tomchick DR, Otomo C et al. Structural basis of actin filament nucleation and processive capping by a formin homology 2 domain. Nature 2005; 433(7025):488-494.
25. Poupel O, Boleti H, Axisa S et al. Toxofilin, a novel actin-binding protein from Toxoplasma gondii, sequesters actin monomers and caps actin filaments. Mol Biol Cell 2000; 11(1):355-368.
26. Klenchin VA, Allingham JS, King R et al. Trisoxazole macrolide toxins mimic the binding of actin-capping proteins to actin. Nat Struct Biol 2003; 10(12):1058-1063.
27. Domanski M, Hertzog M, Coutant J et al. Coupling of folding and binding of thymosin beta4 upon interaction with monomeric actin monitored by nuclear magnetic resonance. J Biol Chem 2004; 279(22):23637-23645.
28. Yarmola EG, Parikh S, Bubb MR. Formation and implications of a ternary complex of profilin, thymosin beta 4, and actin. J Biol Chem 2001; 276(49):45555-45563.
29. Ballweber E, Hannappel E, Huff T et al. Polymerisation of chemically cross-linked actin:thymosin beta(4) complex to filamentous actin: Alteration in helical parameters and visualisation of thymosin beta(4) binding on F-actin. J Mol Biol 2002; 315(4):613-625.
30. McGough A, Pope B, Chiu W et al. Cofilin changes the twist of F-actin: Implications for actin filament dynamics and cellular function. J Cell Biol 1997; 138(4):771-781.
31. Schuler H. ATPase activity and conformational changes in the regulation of actin. Biochim Biophys Acta 2001; 1549(2):137-147.
32. Graceffa P, Dominguez R. Crystal structure of monomeric actin in the ATP state: Structural basis of nucleotide-dependent actin dynamics. J Biol Chem 2003; 278(36):34172-34180.
33. Holmes KC, Geeves MA. The structural basis of muscle contraction. Philos Trans R Soc Lond B Biol Sci 2000; 355(1396):419-431.
34. Hatanaka H, Ogura K, Moriyama K et al. Tertiary structure of destrin and structural similarity between two actin-regulating protein families. Cell 1996; 85(7):1047-1055.
35. Fedorov AA, Lappalainen P, Fedorov EV et al. Structure determination of yeast cofilin. Nat Struct Biol 1997; 4(5):366-369.

36. Leonard SA, Gittis AG, Petrella EC et al. Crystal structure of the actin-binding protein actophorin from Acanthamoeba. Nat Struct Biol 1997; 4(5):369-373.
37. Bowman GD, Nodelman IM, Hong Y et al. A comparative structural analysis of the ADF/cofilin family. Proteins 2000; 41(3):374-384.
38. Paavilainen VO, Merckel MC, Falck S et al. Structural conservation between the actin monomer-binding sites of twinfilin and actin-depolymerizing factor (ADF)/cofilin. J Biol Chem 2002; 277(45):43089-43095.
39. Lappalainen P, Fedorov EV, Fedorov AA et al. Essential functions and actin-binding surfaces of yeast cofilin revealed by systematic mutagenesis. EMBO J 1997; 16(18):5520-5530.
40. Guan JQ, Vorobiev S, Almo SC et al. Mapping the G-actin binding surface of cofilin using synchrotron protein footprinting. Biochemistry 2002; 41(18):5765-5775.
41. Ojala PJ, Paavilainen V, Lappalainen P. Identification of yeast cofilin residues specific for actin monomer and PIP2 binding. Biochemistry 2001; 40(51):15562-15569.
42. Wriggers W, Tang JX, Azuma T et al. Cofilin and gelsolin segment-1: Molecular dynamics simulation and biochemical analysis predict a similar actin binding mode. J Mol Biol 1998; 282(5):921-932.
43. Holmes KC, Popp D, Gebhard W et al. Atomic model of the actin filament. Nature 1990; 347(6288):44-49.
44. Otterbein LR, Graceffa P, Dominguez R. The crystal structure of uncomplexed actin in the ADP state. Science 2001; 293(5530):708-711.
45. Burtnick LD. Modification of actin with fluorescein isothiocyanate. Biochim Biophys Acta 1984; 791(1):57-62.
46. Khaitlina SY, Strzelecka-Golaszewska H. Role of the DNase-I-binding loop in dynamic properties of actin filament. Biophys J 2002; 82(1 Pt 1):321-334.
47. Schwyter DH, Kron SJ, Toyoshima YY et al. Subtilisin cleavage of actin inhibits in vitro sliding movement of actin filaments over myosin. J Cell Biol 1990; 111(2):465-470.
48. Hegyi G, Mak M, Kim E et al. Intrastrand cross-linked actin between Gln-41 and Cys-374. I. Mapping of sites cross-linked in F-actin by N-(4-azido-2-nitrophenyl) putrescine. Biochemistry 1998; 37(51):17784-17792.
49. Kim E, Phillips M, Hegyi G et al. Intrastrand cross-linked actin between Gln-41 and Cys-374. II. Properties of cross-linked oligomers. Biochemistry 1998; 37(51):17793-17800.
50. Holmes KC, Angert I, Kull FJ et al. Electron cryo-microscopy shows how strong binding of to actin releases nucleotide. Nature 2003; 425(6956):423-427.
51. Rayment I, Holden HM, Whittaker M et al. Structure of the actin-myosin complex and its implications for muscle contraction. Science 1993; 261(5117):58-65.
52. McGough A, Way M, DeRosier D. Determination of the alpha-actinin-binding site on actin filaments by cryoelectron microscopy and image analysis. J Cell Biol 1994; 126(2):433-443.
53. Hanein D, Volkmann N, Goldsmith S et al. An atomic model of fimbrin binding to F-actin and its implications for filament crosslinking and regulation. Nat Struct Biol 1998; 5(9):787-792.
54. Sutherland-Smith AJ, Moores CA, Norwood FL et al. An atomic model for actin binding by the CH domains and spectrin-repeat modules of utrophin and dystrophin. J Mol Biol 2003; 329(1):15-33.
55. Moores CA, Keep NH, Kendrick-Jones J. Structure of the utrophin actin-binding domain bound to F-actin reveals binding by an induced fit mechanism. J Mol Biol 2000; 297(2):465-480.
56. Galkin VE, Orlova A, VanLoock MS et al. The utrophin actin-binding domain binds F-actin in two different modes: Implications for the spectrin superfamily of proteins. J Cell Biol 2002; 157(2):243-251.
57. McGough A. F-actin-binding proteins. Curr Opin Struct Biol 1998; 8(2):166-176.

Index